HIGH TIDE IN TUCSON

ALSO BY THE AUTHOR

Fiction
Pigs in Heaven
Animal Dreams
Homeland and Other Stories
The Bean Trees

Poetry
Another America

Nonfiction
Holding the Line: Women in the
Great Arizona Mine Strike of 1983

ESSAYS FROM

NOW OR NEVER

HIGH TIDE IN TUCSON

BARBARA KINGSOLVER

ILLUSTRATIONS BY PAUL MIROCHA

 HarperCollins*Publishers*

FIRST EDITION

Designed by Nancy Singer

Library of Congress Cataloging-in-Publication Data
Kingsolver, Barbara.
    High tide in Tucson : essays from now or never / Barbara Kingsolver.
        p.   cm.
    ISBN 0-06-017291-6
    I. Title
PS3561.I496H54   1995
814' .54—dc20   95-670

95 96 97 98 99 ❖/RRD   10 9 8 7 6 5 4 3 2 1

for Steven,
and every singing miracle

CONTENTS

CONTENTS

PREFACE

When I told my mother I was making a book of my essays, many of which had been published previously in magazines, she responded with pure maternal advocacy: "Oh, good! I think there are some out there that I've missed."

Hurray for Moms, who give us the courage to take up our shelf space on the planet, but I know I can't count on the rest of humanity for the same passion to read every line that ever leaked from my pen. A magazine piece is meant to bloom like an ephemeral flower on the page, here today and recycled tomorrow, but it's another matter to commit those words to acid-free paper and have them skulking on bookshelves for the rest of my natural life. When I began to assemble these pieces, I found that every one begged for substantial revision. Some were magaziney in tone, and needed to be more bookish. Others, when let out of the bag of a journal's tight word limit, grew wild as kudzu vines.

(One of the longest, "Making Peace," began life as three para-
graphs in a doctor's-office magazine.) Then I had to prune
them all back again, and impose unities of theme and tone on
tracts with disparate origins. Most of the essays are now
altered almost beyond recognition since their debut, and
seven are new, written for this collection. My intent was to
make it a *book*, with a beginning, an end, and a modicum of
reason. The essays are meant to be read in order, since some
connect with and depend on their predecessors. However, I
once heard from a reader in Kansas that he always starts
books in the middle—even novels!—so what do I know?

Because so many of the pieces did begin as magazine
articles, the book owes a great deal to all the editors who've
worked with me patiently over the years. I'm particularly
indebted to Paul Trachtman, previously at *Smithsonian
Magazine*, who first talked me into the genre of creative nonfiction; Nancy Newhouse at the *New York Times Magazine*,
who saw me through civil war in Togo, toenail loss in Hawaii,
and more; the staff at *Parenting*, who suggested new angles on
timeless themes, and called back to help after a phone conference I had ended abruptly so as to chase burglars from my
house; the editors at *Natural History*, who invited me back
from poetics to science; Carol Sadtler at *Lands' End*, who
boldly printed my antifashion manifesto in a clothing magazine; Lisa Weinerman at the Nature Conservancy; and the
formidable fact-checking team at the *New York Times*, whose
aptitude for thoroughness will stand them in good stead with
St. Peter—he will, I expect, hire them.

Without the friendship and wise guidance of my literary agent, Frances Goldin, I would still be laboring in a cubicle as a technical writer, and that is the truth. She has taken such risks for me I can hardly count them. So has Janet Goldstein, my bright, faithful star of an editor at HarperCollins. Many other friends and colleagues contributed to this book in different ways, all essential, especially Ann Kingsolver, Joy Johannessen, Janice Bowers, Anne Mairs, Emma Hardesty, Julie Mirocha, and Kelly Brown. Paul Mirocha brought remarkable insights to the work of illustrating the book and remains an inspiring collaborator. (Our first joint project, in 1984, was a document on biological engineering.)

I'm grateful to friends and family members who tolerated or even encouraged my use of our common experience here, to my own ends. I don't think I've let out any secrets (in some cases I've changed names and events just a little, to make sure), but I know it can be jarring to find a piece of yourself stitched into someone else's tale. I cherish the patches contributed by my loved ones, while recognizing they all have their own quilts, too, quite different from mine. I especially thank my lucky stars for Camille Kingsolver, who at this moment is still delighted to see herself drawn in her mother's hand; I apologize now, Camille, for the day you'll feel differently—you have my permission to tell your friends I'm a lunatic and made up every word. And last, and first, eternal thanks to Steven Hopp, my keenest critic and purest enthusiast. Maybe I could have done it alone. But I sure wouldn't want to.

HIGH TIDE IN TUCSON

A hermit crab lives in my house. Here in the desert he's hiding out from local animal ordinances, at minimum, and maybe even the international laws of native-species transport. For sure, he's an outlaw against nature. So be it.

He arrived as a stowaway two Octobers ago. I had spent a week in the Bahamas, and while I was there, wishing my daughter could see those sparkling blue bays and sandy coves, I did exactly what she would have done: I collected shells. Spiky murexes, smooth purple moon shells, ancient-looking whelks sand-blasted by the tide—I tucked them in the pockets of my shirt and shorts until my lumpy, suspect hemlines gave me away, like a refugee smuggling the family fortune. When it was time to go home, I rinsed my loot in the sink and packed it carefully into a plastic carton, then nested it deep in my suitcase for the journey to Arizona.

I got home in the middle of the night, but couldn't wait till morning to show my hand. I set the carton on the coffee table for my daughter to open. In the dark living room her face glowed, in the way of antique stories about children and treasure. With perfect delicacy she laid the shells out on the table, counting, sorting, designating scientific categories like yellow-striped pinky, Barnacle Bill's pocketbook . . . Yeek! She let loose a sudden yelp, dropped her booty, and ran to the far end of the room. The largest, knottiest whelk had begun to move around. First it extended one long red talon of a leg, tap-tap-tapping like a blind man's cane. Then came half a dozen more red legs, plus a pair of eyes on stalks, and a purple claw that snapped open and shut in a way that could not mean We Come in Friendship.

Who could blame this creature? It had fallen asleep to the sound of the Caribbean tide and awakened on a coffee table in Tucson, Arizona, where the nearest standing water source of any real account was the municipal sewage-treatment plant.

With red stiletto legs splayed in all directions, it lunged and jerked its huge shell this way and that, reminding me of the scene I make whenever I'm moved to rearrange the living-room sofa by myself. Then, while we watched in stunned reverence, the strange beast found its bearings and began to reveal a determined, crabby grace. It felt its way to the edge of the table and eased itself over, not falling bang to the floor but hanging suspended underneath within the long grasp of its ice-tong legs, lifting any two or three at a time while many others still held in place. In this remarkable fashion it scrambled around the underside of the table's rim, swift and sure and fearless like a rock climber's dream.

If you ask me, when something extraordinary shows up in your life in the middle of the night, you give it a name and make it the best home you can.

The business of naming involved a grasp of hermit-crab gender that was way out of our league. But our household had a deficit of males, so my daughter and I chose Buster, for balance. We gave him a terrarium with clean gravel and a small cactus plant dug out of the yard and a big cockleshell full of tap water. All this seemed to suit him fine. To my astonishment our local pet store carried a product called Vitaminized Hermit Crab Cakes. Tempting enough (till you read the ingredients) but we passed, since our household leans more toward the recycling ethic. We give him leftovers. Buster's rapture is the day I drag the unidentifiable things in cottage cheese containers out of the back of the fridge.

We've also learned to give him a continually changing assortment of seashells, which he tries on and casts off like Cinderella's stepsisters preening for the ball. He'll sometimes try to squeeze into ludicrous outfits too small to contain him (who can't relate?). In other moods, he will disappear into a conch the size of my two fists and sit for a day, immobilized by the weight of upward mobility. He is in every way the perfect housemate: quiet, entertaining, and willing to eat up the trash. He went to school for first-grade show-and-tell, and was such a hit the principal called up to congratulate me (I think) for being a broad-minded mother.

It was a long time, though, before we began to understand the content of Buster's character. He required more patient observation than we were in the habit of giving to a small, cold-blooded life. As months went by, we would periodically notice with great disappointment that Buster seemed to be dead. Or not entirely dead, but ill, or maybe suffering the crab equivalent of the blues. He would burrow into a gravelly corner, shrink deep into his shell, and not move, for days and days. We'd take him out to play, dunk him in water, offer him a new frock—nothing. He wanted to be still.

Life being what it is, we'd eventually quit prodding our sick friend to cheer up, and would move on to the next stage of a difficult friendship: neglect. We'd ignore him wholesale, only to realize at some point later on that he'd lapsed into hyperactivity. We'd find him ceaselessly patrolling the four corners of his world, turning over rocks, rooting out and dragging around truly disgusting pork-chop bones, digging up his cactus and replanting it on its head. At night when the household fell silent I would lie in bed listening to his methodical pebbly racket from the opposite end of the house. Buster was manic-depressive.

I wondered if he might be responding to the moon. I'm partial to lunar cycles, ever since I learned as a teenager that human females in their natural state—which is to say, sleeping outdoors—arrive at menses in synchrony and ovulate with the full moon. My imagination remains captive to that primordial village: the comradely grumpiness of new-moon days, when the entire world at once would go on PMS alert. And the compensation that would turn up two weeks later on a wild wind, under that great round headlamp, driving both men and women to distraction with the overt prospect of conception. The surface of the land literally rises and falls—as much as fifty centimeters!—as the moon passes over, and we clay-footed mortals fall like dominoes before the swell. It's no surprise at all if a full moon inspires lyricists to corny love songs, or inmates to slamming themselves against barred windows. A hermit crab hardly seems this impetuous, but animals are notoriously responsive to the full moon: wolves howl; roosters announce daybreak all night. Luna moths, Arctic loons, and lunatics have a sole inspiration in common.

Buster's insomniac restlessness seemed likely to be a part of the worldwide full-moon fellowship.

But it wasn't, exactly. The full moon didn't shine on either end of his cycle, the high or the low. We tried to keep track, but it soon became clear: Buster marched to his own drum. The cyclic force that moved him remained as mysterious to us as his true gender and the workings of his crustacean soul.

Buster's aquarium occupies a spot on our kitchen counter right next to the coffeepot, and so it became my habit to begin mornings with chin in hands, pondering the oceanic mysteries while awaiting percolation. Finally, I remembered something. Years ago when I was a graduate student of animal behavior, I passed my days reading about the likes of animals' internal clocks. Temperature, photoperiod, the rise and fall of hormones—all these influences have been teased apart like so many threads from the rope that pulls every creature to its regulated destiny. But one story takes the cake. F. A. Brown, a researcher who is more or less the grandfather of the biological clock, set about in 1954 to track the cycles of intertidal oysters. He scooped his subjects from the clammy coast of Connecticut and moved them into the basement of a laboratory in landlocked Illinois. For the first fifteen days in their new aquariums, the oysters kept right up with their normal intertidal behavior: they spent time shut away in their shells, and time with their mouths wide open, siphoning their briny bath for the plankton that sustained them, as the tides ebbed and flowed on the distant Connecticut shore. In the next two weeks, they made a mystifying shift. They still carried out their cycles in unison, and were regular as the tides, but their high-tide behavior didn't coincide with high tide in Connecticut, or for that matter California, or any other tidal charts known to science. It dawned on the researchers after some calculations that the oysters were responding to high tide in Chicago. Never mind that the gentle mollusks lived in glass boxes in the basement of a steel-and-cement building. Nor that Chicago has no ocean. In the circumstances, the oysters were doing their best.

When Buster is running around for all he's worth, I can only presume it's high tide in Tucson. With or without evidence, I'm romantic enough to believe it. This is the lesson of Buster, the poetry that camps outside the halls of science: Jump for joy, hallelujah. Even a desert has tides.

When I was twenty-two, I donned the shell of a tiny yellow Renault and drove with all I owned from Kentucky to Tucson. I was a typical young American, striking out. I had no earthly notion that I was bringing on myself a calamity of the magnitude of the one that befell poor Buster. I am the commonest kind of North American refugee: I believe I like it here, far-flung from my original home. I've come to love the desert that bristles and breathes and sleeps outside my windows. In the course of seventeen years I've embedded myself in a family here—neighbors, colleagues, friends I can't foresee living without, and a child who is native to this ground, with loves of her own. I'm here for good, it seems.

And yet I never cease to long in my bones for what I left behind. I open my eyes on every new day expecting that a creek will run through my backyard under broad-leafed maples, and that my mother will be whistling in the kitchen. Behind the howl of coyotes, I'm listening for meadowlarks. I sometimes ache to be rocked in the bosom of the blood relations and busybodies of my childhood. Particularly in my years as a mother without a mate, I have deeply missed the safety net of extended family.

In a city of half a million I still really look at every face, anticipating recognition, because I grew up in a town where every face meant something to me. I have trouble remembering to lock the doors. Wariness of strangers I learned the hard way. When I was new to the city, I let a man into my house one hot afternoon because he seemed in dire need of a drink of water; when I turned from the kitchen sink I found sharpened steel shoved against my belly. And so I know, I know. But I cultivate suspicion with as much difficulty as I force tomatoes to grow in the drought-stricken hardpan of my strange backyard. No creek runs here, but I'm still listening to secret tides, living as if I belonged to an earlier place: not Kentucky, necessarily, but a welcoming earth and a human family. A forest. A species.

In my life I've had frightening losses and unfathomable gifts: A knife in my stomach. The death of an unborn child. Sunrise in a rain forest. A stupendous column of blue butterflies rising from a Greek monastery. A car that spontaneously caught fire while I was driving it. The end of a marriage, followed by a year in which I could barely understand how to keep living. The discovery, just weeks ago when I rose from my desk and walked into the kitchen, of three strangers industriously relieving my house of its contents.

I persuaded the strangers to put down the things they were holding (what a bizarre tableau of anti-Magi they made, these three unwise men, bearing a camera, an electric guitar, and a Singer sewing machine), and to leave my home, pronto. My daughter asked excitedly when she got home from school, "Mom, did you say bad words?" (I told her this was the very occasion that bad words exist for.) The police said, variously, that I was lucky, foolhardy, and "a brave lady." But it's not good luck to be invaded, and neither foolish nor brave to stand your ground. It's only the way life goes, and I did it, just as years ago I fought off the knife; mourned the lost child; bore witness to the rain forest; claimed the blue butterflies as Holy Spirit in my private pantheon; got out of the burning car; survived the divorce by putting one foot in front of the other and taking good care of my child. On most important occasions, I cannot think how to respond, I simply do. What does it mean, anyway, to be an animal in human clothing? We carry around these big brains of ours like the crown jewels, but mostly I find that millions of years of evolution have prepared me for one thing only: to follow internal rhythms. To walk upright, to protect my loved ones, to cooperate with my family group—however broadly I care to define it—to do whatever will help us thrive. Obviously, some habits that saw us through the millennia are proving hazardous in a modern context: for example, the yen to consume carbohydrates and fat whenever they cross our path, or the proclivity for unchecked reproduction. But it's surely worth forgiving ourselves these tendencies a little, in light of the fact that they are what got us here. Like Buster, we are creatures of inexplicable cravings. Thinking isn't everything. The way I stock my refrigerator would amuse a level-headed interplanetary observer, who would see I'm responding not to real necessity but to the dread of famine honed in the African savannah. I can laugh at my Rhodesian Ridgeback as she furtively sniffs the houseplants for a place to bury bones, and circles to beat down the grass before lying on my kitchen floor. But she and I are exactly the same kind of hairpin.

We humans have to grant the presence of some past adaptations, even in their unforgivable extremes, if only to admit they are permanent rocks in the stream we're obliged to navigate. It's easy to speculate and hard to prove, ever, that genes control our behaviors. Yet we are persistently, excruciatingly adept at many things that seem no more useful to modern life than the tracking of tides in a desert. At recognizing insider/outsider status, for example, starting with white vs. black and grading straight into distinctions so fine as to baffle the bystander—Serb and Bosnian, Hutu and Tutsi, Crip and Blood. We hold that children learn discrimination from their parents, but they learn it fiercely and well, world without end. Recite it by rote like a multiplication table. Take it to heart, though it's neither helpful nor appropriate, anymore than it is to hire the taller of two men applying for a position as bank clerk, though statistically we're likely to do that too. Deference to the physical superlative, a preference for the scent of our own clan: a thousand anachronisms dance down the strands of our DNA from a hidebound tribal past, guiding us toward the glories of survival, and some vainglories as well. If we resent being bound by these ropes, the best hope is to seize them out like snakes, by the throat, look them in the eye and own up to their venom.

But we rarely do, silly egghead of a species that we are. We invent the most outlandish intellectual grounds to justify discrimination. We tap our toes to chaste love songs about the silvery moon without recognizing them as hymns to copulation. We can dress up our drives, put them in three-piece suits or ballet slippers, but still they drive us. The wonder of it is that our culture attaches almost unequivocal shame to our animal nature, believing brute urges must be hurtful, violent things. But it's no less an animal instinct that leads us to marry (species that benefit from monogamy tend to practice it); to organize a neighborhood cleanup campaign (rare and doomed is the creature that fouls its nest); to improvise and enforce morality (many primates socialize their young to be cooperative and ostracize adults who won't share food).

It's starting to look as if the most shameful tradition of Western civilization is our need to deny we are animals. In just a few centuries of setting ourselves apart as landlords of the Garden of Eden, exempt from the natural order and entitled to hold dominion, we have managed to behave like so-called animals anyway, and on top of it to wreck most of what took three billion years to assemble. Air, water, earth, and fire—so much of our own element so vastly contaminated, we endanger our own future. Apparently we never owned the place after all. Like every other animal, we're locked into our niche: the mercury in the ocean, the pesticides on the soybean fields, all come home to our breast-fed babies. In the silent spring we are learning it's easier to escape from a chain gang than a food chain. Possibly we will have the sense to begin a new century by renewing our membership in the Animal Kingdom.

Not long ago I went backpacking in the Eagle Tail Mountains. This range is a trackless wilderness in western Arizona that most people would call Godforsaken, taking for granted God's preference for loamy topsoil and regular precipitation. Whoever created the Eagle Tails had dry heat on the agenda, and a thing for volcanic rock. Also cactus, twisted mesquites, and five-alarm sunsets. The hiker's program in a desert like this is dire and blunt: carry in enough water to keep you alive till you can find a water source; then fill your bottles and head for the next one, or straight back out. Experts warn adventurers in this region, without irony, to drink their water while they're still alive, as it won't help later.

Several canyons looked promising for springs on our topographical map, but turned up dry. Finally, at the top of a narrow, overgrown gorge we found a blessed tinaja, a deep, shaded hollow in the rock about the size of four or five claw-foot tubs, holding water. After we drank our fill, my friends struck out again, but I opted to stay and spend the day in the hospitable place that had slaked our thirst. On either side of the natural water tank, two shallow caves in the canyon wall faced each other, only a few dozen steps apart. By crossing from one to the other at noon, a person could spend the whole day here in shady comfort—or in colder weather, follow the winter sun. Anticipating a morning of reading, I pulled *Angle of Repose* out of my pack and looked for a place to settle on the flat, dusty floor of the west-facing shelter. Instead, my eyes were startled by a smooth corn-grinding stone. It sat in the exact center of its rock bowl, as if the Hohokam woman or man who used this mortar and pestle had walked off and left them there an hour ago. The Hohokam disappeared from the earth in A.D. 1450. It was inconceivable to me that no one had been here since then, but that may have been the case—that is the point of trackless wilderness. I picked up the grinding stone. The size and weight and smooth, balanced perfection of it in my hand filled me at once with a longing to possess it. In its time, this excellent stone was the most treasured thing in a life, a family, maybe the whole neighborhood. To whom it still belonged. I replaced it in the rock depression, which also felt smooth to my touch. Because my eyes now understood how to look at it, the ground under my feet came alive with worked flint chips and pottery shards. I walked across to the other cave and found its floor just as lively with historic debris. Hidden under brittlebush and catclaw I found another grinding stone, this one some distance from the depression in the cave floor that once answered its pressure daily, for the grinding of corn or mesquite beans.

For a whole day I marveled at this place, running my fingers over the knife edges of dark flint chips, trying to fit together thick red pieces of shattered clay jars, biting my lower lip like a child concentrating on a puzzle. I tried to guess the size of whole pots from the curve of the broken pieces: some seemed as small as my two cupped hands, and some maybe as big as a bucket. The sun scorched my neck, reminding me to follow the shade across to the other shelter. Bees hummed at the edge of the water hole, nosing up to the water, their abdomens pulsing like tiny hydraulic pumps; by late afternoon they rimmed the pool completely, a collar of busy lace. Off and on, the lazy hand of a hot breeze shuffled the white leaves of the brittlebush. Once I looked up to see a screaming pair of red-tailed hawks mating in midair, and once a clatter of hooves warned me to hold still. A bighorn ram emerged through the brush, his head bent low under his hefty cornice, and ambled by me with nothing on his mind so much as a cool drink.

How long can a pestle stone lie still in the center of its mortar? That long ago—that recently—people lived here. *Here*, exactly, and not one valley over, or two, or twelve, because this place had all a person needs: shelter, food, and permanent water. They organized their lives around a catchment basin in a granite boulder, conforming their desires to the earth's charities; they never expected the opposite. The stories I grew up with lauded Moses for striking the rock and bringing forth the bubbling stream. But the stories of the Hohokam—oh, how they must have praised that good rock.

At dusk my friends returned with wonderful tales of the ground they had covered. We camped for the night, refilled our canteens, and hiked back to the land of plumbing and a fair guarantee of longevity. But I treasure my memory of the day I lingered near water and covered no ground. I can't think of a day in my life in which I've had such a clear fix on what it means to be human.

Want is a thing that unfurls unbidden like fungus, opening large upon itself, stopless, filling the sky. But *needs*, from one day to the next, are few enough to fit in a bucket, with room enough left to rattle like brittlebush in a dry wind.

For each of us—furred, feathered, or skinned alive—the whole earth balances on the single precarious point of our own survival. In the best of times, I hold in mind the need to care for things beyond the self: poetry, humanity, grace. In other times, when it seems difficult merely to survive and be happy about it, the condition of my thought tastes as simple as this: let me be a good animal today. I've spent months at a stretch, even years, with that taste in my mouth, and have found that it serves.

But it seems a wide gulf to cross, from the raw, green passion for survival to the dispassionate, considered state of human grace. How does the animal mind construct a poetry for the modern artifice in which we now reside? Often I feel as disoriented as poor Buster, unprepared for the life that zooms headlong past my line of sight. This clutter of human paraphernalia and counterfeit necessities—what does it have to do with the genuine business of life on earth? It feels strange to me to be living in a box, hiding from the steadying influence of the moon; wearing the hide of a cow, which is supposed to be dyed to match God-knows-what, on my feet; making promises over the telephone about things I will do at a precise hour next *year*. (I always feel the urge to add, as my grandmother does, "Lord willing and the creeks don't rise!") I find it impossible to think, with a straight face, about what colors ought not to be worn after Labor Day. I can become hysterical over the fact that someone, somewhere, invented a thing called the mushroom scrubber, and that many other people undoubtedly feel they *need* to possess one. It's completely usual for me to get up in the morning, take a look around, and laugh out loud.

Strangest of all, I am carrying on with all of this in a desert, two thousand miles from my verdant childhood home. I am disembodied. No one here remembers how I was before I grew to my present height. I'm called upon to reinvent my own childhood time and again; in the process, I wonder how I can ever know the truth about who I am. If someone had told me what I was headed for in that little Renault—that I was stowing away in a shell, bound to wake up to an alien life on a persistently foreign shore—I surely would not have done it. But no one warned me. My culture, as I understand it, values independence above all things—in part to ensure a mobile labor force, grease for the machine of a capitalist economy. Our fairy tale commands: Little Pig, go out and seek your fortune! So I did.

Many years ago I read that the Tohono O'odham, who dwell in the deserts near here, traditionally bury the umbilicus of a newborn son or daughter somewhere close to home and plant a tree over it, to hold the child in place. In a sentimental frame of mind, I did the same when my own baby's cord fell off. I'm staring at the tree right now, as I write—a lovely thing grown huge outside my window, home to woodpeckers, its boughs overarching the house, as dissimilar from the sapling I planted seven years ago as my present life is from the tidy future I'd mapped out for us all when my baby was born. She will roam light-years from the base of that tree. I have no doubt of it. I can only hope she's growing as the tree is, absorbing strength and rhythms and a trust in the seasons, so she will always be able to listen for home.

I feel remorse about Buster's monumental relocation; it's a weighty responsibility to have thrown someone else's life into permanent chaos. But as for my own, I can't be sorry I made the trip. Most of what I learned in the old place seems to suffice for the new: if the seasons like Chicago tides come at ridiculous times and I have to plant in September instead of May, and if I have to make up family from scratch, what matters is that I do have sisters and tomato plants, the essential things. Like Buster, I'm inclined to see the material backdrop of my life as mostly immaterial, compared with what moves inside of me. I hold on to my adopted shore, chanting private vows: wherever I am, let me never forget to distinguish *want* from *need*. Let me be a good animal today. Let me dance in the waves of my private tide, the habits of survival and love.

Every one of us is called upon, probably many times, to start a new life. A frightening diagnosis, a marriage, a move, loss of a job or a limb or a loved one, a graduation, bringing a new baby home: it's impossible to think at first how this all will be possible. Eventually, what moves it all forward is the subterranean ebb and flow of being alive among the living.

In my own worst seasons I've come back from the colorless world of despair by forcing myself to look hard, for a long time, at a single glorious thing: a flame of red geranium outside my bedroom window. And then another: my daughter in a yellow dress. And another: the perfect outline of a full, dark sphere behind the crescent moon. Until I learned to be in love with my life again. Like a stroke victim retraining new parts of the brain to grasp lost skills, I have taught myself joy, over and over again.

It's not such a wide gulf to cross, then, from survival to poetry. We hold fast to the old passions of endurance that buckle and creak beneath us, dovetailed, tight as a good wooden boat to carry us onward. And onward full tilt we go, pitched and wrecked and absurdly resolute, driven in spite of everything to make good on a new shore. To be hopeful, to embrace one possibility after another—that is surely the basic instinct. Baser even than hate, the thing with teeth, which can be stilled with a tone of voice or stunned by beauty. If the whole world of the living has to turn on the single point of remaining alive, that pointed endurance is the poetry of hope. The thing with feathers.

What a stroke of luck. What a singular brute feat of outrageous fortune: to be born to citizenship in the Animal Kingdom. We love and we lose, go back to the start and do it right over again. For every heavy forebrain solemnly cataloging the facts of a harsh landscape, there's a rush of intuition behind it crying out: High tide! Time to move out into the glorious debris. Time to take this life for what it is.

CREATION STORIES

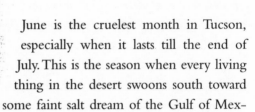

June is the cruelest month in Tucson, especially when it lasts till the end of July. This is the season when every living thing in the desert swoons south toward some faint salt dream of the Gulf of Mexico: tasting the horizon, waiting for the summer storms. This year they are late. The birds are pacing the ground stiff-legged, panting, and so am I. Waiting. In this blind, bright still-June weather the shrill of the cicadas hurts your eyes. Every plant looks pitiful and, when you walk past it, moans a little, envious because you can walk yourself to a drink and it can't.

The water that came last winter is long gone. "Female rain," it's called in Navajo: the gentle, furtive rains that fall from overcast skies between November and March. That was weather to drink and to grow on. But not to remember, anymore than a child remembers last birthday's ice cream, once the months have passed without another drop. In June there is no vital sign, not so much as a humid breath against a pane of glass, till the summer storms arrive. What we're waiting for now is male rain. Big, booming wait-till-your-father-gets-home cloudbursts that bully up from Mexico and threaten to rip the sky.

The Tohono O'odham have lived in the Sonoran Desert longer than anyone else who's still living; their answer to this season is to make frothy wine from the ripe saguaro fruits, and drink it all day and all night in a do-or-die ceremony to bring down the first storm. When it comes, the answer to a desert's one permanent question, that first storm defines the beginning of the Tohono O'odham new year. The storms themselves are enough to get drunk on: ferocious thunder and raindrops splatting so hard on the cooked ground you hear the thing approaching like mortar fire.

I saw my first of these summer storms in 1978. I hadn't been in Arizona long enough to see the calendar open and close, so I spent the early summer in a state of near panic, as the earliest people in any place must have done when they touched falling snow or the dry season's dust and asked each time: This burning cold, these dying plants—is this, then, the end of the world?

I lived in a little stuccoed house in a neighborhood of barking dogs and front-yard shrines to the Virgin of Guadalupe. One sweltering afternoon I heard what I believed must be kids throwing gravel at the houses, relentlessly and with feeling. It was hot enough so that the neighborhood, all of it, dogs and broken glass on the sidewalks included, had murder in mind. I knew I was risking my neck to go outside and scold kids for throwing rocks, but I went anyway. What I saw from the front stoop arrested me in my footprints: not a troop of juvenile delinquents, but a black sky and a wall of water as high as heaven, moving up the block. I ran into the street barefoot and danced with my mouth open. So did half my neighbors. Armistice Day.

Now I live on the outskirts of town, in the desert at the foot of the Tucson Mountains, where waiting for the end of the drought becomes an obsession. It's literally 110 degrees in the shade today, the kind of weather real southwesterners love to talk about. We have our own kind of Jack London thing, in reverse: Remember that year (swagger, thumbs in the belt) when it was 122 degrees and planes couldn't land at the airport?

This is actually true. For years I held the colorful impression that the tarmac had liquefied, so that aircraft would have plowed into it like mammoth flies bellying into ointment. Eventually an engineer gave me a pedestrian, probably accurate, explanation about heat interfering with the generation of lift above the wings. Either way, weather that stops modern air traffic is high drama in America.

We revel in our misery only because we know the end, when it comes, is so good. One day there will be a crackling, clean, creosote smell in the air and the ground will be charged and the hair on your arms will stand on end and then BOOM, you are thrillingly drenched. All the desert toads crawl out of their burrows, swell out their throats, and scream for sex while the puddles last. The ocotillos leaf out before your eyes, like a nature show on fast forward. There is so little time before the water sizzles back to thin air again. So little time to live a whole life in the desert. This is elemental mortality, the root of all passion.

Since I moved to this neighborhood of desert, I've learned I have other writers for neighbors. Unlike the toads, we're shy—we don't advertise our presence to each other quite so ostentatiously. In fact, I only found out I'd joined a literary commune when my UPS man—I fancy him a sort of manly Dorothy Parker in uniform—began giving me weekly updates. Visitors up at Silko's had been out looking for wild pigs, and Mr. Abbey had gone out in his backyard and shot the TV, again. (Sad to say, that doesn't happen anymore. We all miss Ed.)

I imagine other neighbors: that Georgia O'Keeffe, for example, is out there walking the hills in sturdy shoes, staring down the UPS man with such a fierce eye that he will never dare tell.

What is it that draws creators to this place? Low rent, I tell my friends who ask, but it's more than that. It's the Southwest: a prickly land where mountain lions make bets with rabbits, and rabbits can win. Where nature rubs belly to belly with subdivision and barrio, and coyotes take shortcuts through the back alleys. Here even the rain has gender, the frogs sing *carpe diem*, and fast teenage girls genuflect quickly toward the door of the church, hedging their bets, as they walk to school in tight skirts and shiny high heels.

When I drive to the post office every few days to pick up my mail, it's only about twelve miles round trip, but I pass through at least half-a-dozen neighborhoods that distinguish themselves one from the other by architecture and language and even, especially, creation myth. First among them is the neighborhood of jackrabbits and saguaros, who imperiously tolerate my home, though I can't speak their language or quite understand their myths.

Then, just inside the city limits, a red cobble of just-alike roofs—paved air—where long strands of exurban condominiums shelter immigrants from Wisconsin, maybe, or Kansas, who dream in green and hug small irrigated lawns to their front doors.

Next I cross the bridge over the Santa Cruz, whose creation story bubbles from ephemeral springs in the mountains of southern Arizona and Mexico. In these lean days she's a great blank channel of sand, but we call her a river anyway, and say it with a straight face too, because in her moods this saint has taken out bridges and houses and people who loved their lives.

Then I pass under the artery of Interstate 10, which originates in Los Angeles or Jacksonville, Florida, depending on your view of destiny; and the railroad track, whose legend is a tale tasting of dynamite, the lives and deaths of immigrants who united a continent and divided in twain the one great original herd of American bison.

Then without warning I am smack in the middle of a Yaqui village that is fringe-edged and small like a postage stamp, and every bit alive. Despite its size, Pascua Yaqui is a sovereign world; I come here every Easter to watch an irresistible pageant combining deer dances with crucifixion. Like the Tohono O'odham singing down the rain, the masked Yaqui dancers listen for the heartbeat of creation, and keep a promise with every vernal equinox to hold the world to its rightful position. On this small patch of dusty ground, the religion of personal salvation is eclipsed by a faith whose question and answer are matters of order in the universe. Religion of that kind can crack your mind open the way lightning splits a pine, leaving the wind to howl through the scorched divide. I can hardly ever even drive through here, in my serviceable old Toyota, without biting my lip and considering immensity.

Calle Ventura marks the entrance to another state, where on a fine, still day your nose can compare the goods from three tortilla factories. From here the sidewalks roll, the walls crumble and shout with territorial inscription, brown dogs lie under cherry Camaros and the Virgin of Guadalupe holds court in the parking lot of the Casa Rey apartments.

Across the street stands the post office, neutral territory: mailboxes all identical, regardless of the keyholder's surname, as physically uniform as a table of contents. We are all equals in the eyes of the USPO, containing our secrets. I grab mine and scuttle away. The trip home takes me right back through all these lands again, all these creation stories, and that's enough culture for one day, usually.

I close the door, breathless, and stare out my window at a landscape of wonders thrown together with no more thought than a rainstorm or a volcano can invoke on its own behalf. It's exactly as John Muir said, as if "nature in wildest extravagance held her bravest structures as common as gravel-piles."

From here I begin my story. I can't think of another place like it.

MAKING PEACE

When I left downtown Tucson to make my home in the desert, I went, like Thoreau, "to live deliberately." I think by this he meant he was tired of his neighbors. For me the problem wasn't specifically my neighbors, whom I loved (and it's a good thing, since our houses were so close together we could lean out our bedroom windows and shake hands), but the kids who spilled over from—and as far as I could see, never actually attended—the high school across the street. They liked rearranging the flowers in my front yard, upside down. They had art contests on my front walk, the point being to see whether a realistic rendition of the male sex organ could be made to span the full sweep from sidewalk to front door. They held very loud celebrations, daily, on my front porch. When my brain was jangled to the limits of reason, I would creep from my writing desk to the front door, poke my head out, and ask if they could turn the music down. They glared, with So What eyes. Informed me this was a party, and I wasn't invited.

The school's principal claimed that kids outside the school grounds were beyond his jurisdiction; I was loath to call the city police, but did (only after the porch party ratified a new sport involving urination), and they told me what I knew they'd say: the principal ought to get those kids in school. My territory was up for grabs, by anyone but me.

After some years had passed and nobody seemed to be graduating, I struck out for Walden. My husband and I sold our house, collected our nerve, and bought four acres of rolling desert—a brambly lap robe thrown over the knees of the Tucson Mountains, a stone's throw beyond the city limits. There was a tiny cabin, which we could expand to suit our needs. I anticipated peace.

Like a pioneer claiming her little plot of prairie, I immediately planted a kitchen garden and hollyhocks outside the door. I inhaled silence, ecstatic with the prospect of owning a place that was really my own: rugged terrain, green with mesquite woods and rich in wildlife. No giant penises waiting to impale me when I threw open my front door. Only giant saguaros. Only bird song and faint hoofprints in the soil, evidence of wild creatures who might pass this way under cover of darkness.

Sure enough they came, the very first night: the javelinas. Woolly pigs. They are peccaries, technically, cloven-hoofed rooters of the New World, native to this soil for much longer than humans have known it—but for all the world they are pigs. I pressed my face to the window when I heard their thumping and rustling. Their black fur bristled as they bumped against one another and snuffled the ground with long, tusked snouts. I watched them eat my hollyhocks one by one.

Pioneering takes patience. I thought maybe that first visit was some kind of animal welcome-wagon tradition in reverse, and that over time we could reach an accord. Night after night, they returned. The accord seemed to be: You plant, we eat. The jackrabbits were hungry too, but I discovered that they shun the nightshade family—which conveniently includes tomatoes and eggplants—and that I could dissuade them from my flowers with chicken wire (although a flowerbed that looks like Fort Knox is a doubtful ornament). Not so picky, the pigs. With mouths of steel and cast-iron stomachs, they relished the nightshades, and in their eagerness I swear they even ate chicken wire. Over the weeks I tried the most pungent flowers I could think of: geraniums, marigolds. They ate everything. Rare is the epicurean pig who has feasted at such a varied table as the one I provided.

I tried to drive them off. Banged on the windows, shrieked, and after a goodly amount of accomplishing nothing whatsoever through those means, cautiously opened the door a crack, stuck my head out, and hollered.

"Shoo, pigs!" said I.

"Not by the hair of my chinny chin chin," thought they, apparently, in what passes for thought within those bony skulls. They ignored me profoundly, inciting me to extremes. I stooped to throwing rocks, and once by the wildest of chances, so help me God, I hit one, broadside. With a rock the size of a softball, and a respectable thud. The victim paused for half an instant midgobble and sniffed the air as if to ask, Was that a change in the weather? Then returned to the hollyhocks at hand. On the He-Man Scale of Strength, my direct hit scored "Weenie." I seethed between the four walls of my house like Rochester's mad wife in the attic.

In a fit of spite I went to a nursery that specializes in exotics, and brought home an *Adenium obesum*. This is the beautiful plant whose singularly lethal sap is used by African hunters to poison their darts.

Javelinas understand spite: they uprooted my *Adenium obesum*, gored it, and left it for dead.

Over the months our house slowly grew, with javelinas watching. We framed up an extra room, which we would eventually connect to the old house by tearing out a window, once it was sealed to the outside. We laid out sheet-metal ductwork, which would go into the ceiling, for heating the new addition. In the middle of the night we woke to the sound of the devil's own celebration: hellacious hoofs on tin drums. The pigs had found their way into the new room and were trampling the ductwork, sending their tinny war cry to the stars above.

Ownership is an entirely human construct. At some point people got along without it. Many theorists have addressed the question of how private property came about, and some have gone so far as to suggest this artificial notion has led us into a mess of trouble. They aren't talking about *personal property*, like a toothbrush or a digging stick to call one's own, which has probably always been a human tradition. Even a bird, after all, has its nest, and chimpanzees in a part of central Africa where there's a scarcity of nut-smashing tools are known to get possessive about their favorite rocks. But to own land, plants, other animals, more stuff than we need—that is the peculiar product of a modern imagination.

In the beginning, humans were communal and social creatures; this is agreed upon by all scientists who've given our species retroactive study. The habit and necessity of cooperation is what led us, like other social species, toward the development of an elaborate communication system. Other social primates that live in large groups, like Japanese macaques and baboons, communicate with a much richer repertoire of sounds than the solitary primates like orangutans. Many social mammals use not only verbal but olfactory signals—a language of the nose. An example of complex communication among birds, familiar to any rural child, is that of the socially cooperative chickens, who use different calls (in the wild, as well as the barnyard) to refer to important events in their lives: krk krk krk (food over here); kark kark KARK (*really good* food over here); RRRR-rrrr (hawk overhead). Parrots, another famous category of garrulous birds, are presumed by scientists to have developed their gift of gab because of social habits and longevity in the wild.

It's safe to presume that the most talkative of all primates, *Homo sapiens*, evolved in the context of cooperative social groups also, hunting and gathering on the African savannahs. The theory that has percolated best into popular imagination is the one that claims men clobbered the animals, providing intermittent jubilations of protein for the home crowd, while women dug roots, picked fruits and seeds, and harvested edible plant parts. The latter activities presumably would provide the bulk of the steady calories, but for many decades the burgeoning science of human origins was captivated by the hunting scenario: the need to peer out over the savannah grass as incentive for walking upright; the necessities of spear making and cooperative hunting giving rise to language, dexterity, and a large, complex brain.

This neat boy-girl theory smacks of sexist backward projection, I've always thought, while I do concede (having carried a toddler on my own hip for a few years) that it's more feasible to go berry picking than lion hunting with a nursing child in tow.

But many early anthropologists, unable to resist drama, apparently overestimated the importance of "the hunt" as a shaper of our body, character, and destiny. It's now understood that the earliest evidence of meat eating in the human archaeological record comes from East African sites that are less than two million years old. Considering that we have been walking upright and approximately human for more than twice that long, carnivory may have been an afterthought. Anthropologist Adrienne Zihlman argues that the challenge that shaped us was most likely the savannah environment itself, which is not a monoculture of tall grass but a complex mosaic of grassland, hills, and forested areas along watercourses. Potential food sources were abundant but seasonal and widely scattered; the early human's home range would have been much larger than that of living savannah baboons and chimpanzees. The best survivors would be those with a good locomotor system and the capacity to carry water and food, as well as offspring. Based on the fossil record, and on close study of living hunter-gatherers and our primate relatives in similar habitats, Zihlman has estimated that plant foods, insects, and small vertebrates made up more than 90 percent of the early hominid diet, and that "scavenging and consumption of large dead animals found by chance" was probably infrequent. This scenario, which has our ancestors shooing off hyenas and vultures from the *carcass du jour,* isn't going to sell any movie rights, but it has the advantage of evidence behind it.

In any case, the best perspective on the notion of a natural division of labor was given me long ago by one of my most influential college professors, Preston Adams, a botanist who studied human evolution. He pointed out that all "man the hunter" theories implicitly establish women as the first botanists. He also liked to tell restless zoology majors that it takes a superlative mind to appreciate a plant. He kindly allowed me to put two and two together.

When it began to dawn on our insightful ancestors that they could save some edible seeds, put them in the ground, and have a whole new edible crop right on the front stoop, we had agriculture on our hands. It's a giant step, the historical materialists maintain, to go from appropriating the products of nature to increasing their supply through human labor. The first evidence of cultivated grains comes from archaeological sites that are in the neighborhood of eleven thousand years old. Joseph Campbell, in his *Atlas of World Mythology,* identifies at least three independent points of origin for "The Way of the Seeded Earth": the Middle East, Southeast Asia, and Central America. Domestication of animals followed right along. A handful of seeds, like Jack's magic beans, turned our fortunes head over heels.

Friedrich Engels, the nineteenth-century economist and close associate of Karl Marx, examined our history under the bright lamp of a new paradigm set forth by his contemporary Charles Darwin. Engels also had access to the prodigious work of anthropologist Lewis Henry Morgan. Countless modern scholars have addressed the history of private property, but it's hard to beat the elegance of Engels's simple outline of human social evolution, laid out in his wonderful classic, *The Origin of the Family, Private Property, and the State.* In the natural progression to a more controlled form of hunting and gathering, he theorized, the community efforts of planting and harvesting remained the female domain, while animals that could "belong" to someone belonged to men. Goats and sheep, being mobile and tradable, became currency. Rather suddenly men got the purse strings. Rather suddenly "purse strings" was a concept. So was "inheritance." The family tightened its boundaries, the better to serve as conduit for property passed from father to son.

If we can divine religion from relics, it seems pretty clear that up to this point human societies stood most in awe of female power: the pregnant Venus of Willendorf; the Woman with the Horn carved on a cliff in Dordogne, France; the fecund clay figurines that preclassical Mexicans buried with their dead; pregnant torsos carved from the tusks of woolly mammoths in Asia; the pale stone fertility figures strewed along the Mediterranean coast like so many dragons' teeth. The one that gets my vote for blunt reverence is a mammoth-ivory disk from a gravesite in Moravia, cut with a single, unambiguous vulval slit. So many goddesses, so little time—for they fell, and fell *far*, from grace. It's pretty difficult now even to imagine female body parts as sacrament: when the kids spray-painted vulvas on my front steps, their thoughts were oh so far from God.

How fiercely doth the sacred turn profane. Our ancestors in the Fertile Crescent appear to have dropped Goddess Mother like a hot rock, and shifted their allegiance to God the Father, coincident with the rise of Man the Owner of the Flock.

Since then, most of us have come to see human ownership of places and things, even other living creatures, as a natural condition, right as rain. While rights and authority and questions of distribution are fiercely debated, the basic concept is rarely in doubt. I remember arguing tearfully, as a child, that a person couldn't own a tree, and still in my heart I believe that, but inevitably to come of age is to own. When we stand upon the ground, we first think to ask, Whose ground is this? And NO TRESPASSING doesn't just mean, "Don't build your house here." It means: "All you see before you, the trees, the songbirds, the poison ivy, the water beneath the ground, the air you would breathe if you passed through here, the grass you would tread upon, the *very idea of existing in this place*—all these are mine." Nought but a human mind could think of such a thing. And nought but a human believes it. Javelinas, and teenagers, still hark to the earth's primordial state and the music of the open range.

Now, territoriality is a different matter. Birds do that. Dogs do it. Pupfish in their little corner of a mud puddle do it. They (meaning, usually, the males of territorial species) mark out a little plot and defend it from others of their own kind, for the duration of their breeding season. This is about reproduction: he is making jolly well sure that any eggs that get fertilized, or babies that get raised, within that hallowed territory are, in fact, his own. Often, it's also a matter of securing an area that contains enough resources—nuts, berries, caterpillars, flower nectar, whatever—to raise a brood of young. Just enough, usually, and hardly a caterpillar more. The minute the young have flown away, the ephemeral territory vanishes back into the thin air, or the bird brain, whence it came. The male might return to establish a breeding territory in the same place again next year, or he might not. The landscape lives on, fairly untouched by the process.

When a male bird—a vireo, for example—sings his belligerent song at another male vireo that approaches his neck of the woods, he is singing about family. It's a little bit like grumbling over the handsome delivery person who's getting too friendly with your spouse; a *lot* like coming with a crowbar after an intruder at your child's bedroom window in the night; and nothing at all like a NO TRESPASSING sign. The vireo doesn't waste his breath on the groundhogs gathering chestnuts under his nose, or the walnut trees using the sunlight to make their food, the grubs churning leaves into soil, the browsing deer, or even other birds that come to glean seeds that are useless to a vireo's children. Worm-eating birds have no truck with seedeaters; small-seed eaters ignore big-seed eaters. This is the marvelous construct of "niche," the very particular way an organism uses its habitat, and it allows for an almost incomprehensible degree of peaceful coexistence. Choose a cubic foot of earth, about anywhere that isn't paved; look closely enough, and you'll find that thousands of different kinds of living things are sharing that place, each one merrily surviving on something its neighbors couldn't use for all the tea in China. I'm told that nine-tenths of human law is about possession. But it seems to me we don't know the first thing about it.

It did not take me long in the desert to realize I was thinking like a person, and on that score was deeply outnumbered. My neighbors weren't into the idea of private property, and weren't interested in learning about it, either. As Kafka frankly put it, when it's you against the world, bet on the world.

So I dispensed with lordship, and went for territoriality. I turned a realistic eye on my needs. I don't really have to have hollyhocks outside my door. But I'd like some tomatoes and eggplants. Oak-leaf lettuce on crisp fall days, and in the spring green beans and snowpeas. Maybe a *little* bed of snapdragons. It wouldn't take much. Since I had no plans to raise a huge brood, sixty square feet or so of garden space would serve me very well.

I revised my blueprints and looked hard at Pueblo architecture, which shuns the monumental for the more enduring value of blending in. The Pueblo, as I understand their way of life, seem to be more territorial than proprietary, and they've lived in the desert for eight centuries. Between the javelinas and me it had come down to poison darts in about eight days. Enough with that.

I settled on a fairly ancient design. The wings of my house

enfold a smallish courtyard. My territorial vireo song is a block wall, eight feet high. Inside the courtyard I grow a vegetable garden, a few fruit trees, and a bright flag of flowerbed that changes its colors every season. The acres that lie beyond the wall I have left to cactus and mesquite bramble, and the appetites that rise to its sharp occasion.

Life is easier since I abdicated the throne. What a relief, to relinquish ownership of unownable things. Engels remarked at the end of his treatise that the outgrowth of property has become so unmanageable that "the human mind stands bewildered in the presence of its own creation." But he continues on a hopeful note: "The time which has passed since civilization began is but a fragment of the past duration of man's existence; and but a fragment of the ages yet to come. . . . A mere property career is not the final destiny of mankind."

Indeed. We're striving hard to get beyond mere property career around here. I've quit with the *Adenium obesum*, and taken to leaving out table scraps for the pigs. I toss, they eat. I find, now that I'm not engaged in the project of despising them, they are rather a hoot to watch. On tiny hooves as preposterous as high-heeled pumps on a pirate, they come mincing up the path. They feel their way through the world with flattened, prehensile snoots that flare like a suction-cup dart, and swivel about for input like radar dishes. When mildly aroused (which is as far as it goes, in the emotional color scheme of the javelina), their spiky fur levitates into a bristly, spherical crown—Tina Turner laced with porcupine. I don't even mind that they come and eat up our jack-o'-lanterns at Halloween; it's worth it. They slay me every time with their hilarious habit of going down on their foreknees and walking along, pious supplicants in awe of life's bounty, pushing whole pumpkins before them.

Meanwhile, in the cloistered territory of the courtyard, so many things come and go it would feel absurd to call it mine: I've seen an elf owl picking through the compost pile; Gila woodpeckers fighting over the tree trunks; hummingbirds at the flowers; doves who nested in the grape arbor; a roadrunner who chased off the doves and gulped down their eggs; a pair of cardinals and a Pyrrhuloxia couple who nested in adjacent trees and became so confused, when the young fledged and flew to the ground, that they hopped around frantically for a week feeding each other's kids. A pair of Swainson's thrushes stopped in for a day on their migratory flight from Canada to Peru; to them, this small lush square in a desert state must have appeared as Moses' freshet from the rock.

The cardinals, of course, eat the grapes. In some years the finches peck a hole in every single apricot before I get around to throwing a net over the tree. A fat, clairvoyant rock squirrel scales the wall and grabs just about every third tomato, on the morning I decide that tomorrow it will be ripe enough to pick.

So what, they all declare with glittering eyes. This is their party, and I wasn't exactly invited.

IN CASE YOU EVER WANT TO GO HOME AGAIN

I have been gone from Kentucky a long time. Twenty years have done to my hill accent what the washing machine does to my jeans: taken out the color and starch, so gradually that I never marked the loss. Something like that has happened to my memories, too, particularly of the places and people I can't go back and visit because they are gone. The ancient brick building that was my grade school, for example, and both my grandfathers. They're snapshots of memory for me now, of equivocal focus, loaded with emotion, undisturbed by anyone else's idea of the truth. The schoolhouse's plaster ceilings are charted with craters like maps of the moon and likely to crash down without warning. The windows are watery, bubbly glass reinforced with chicken wire. The weary wooden staircases, worn shiny smooth in a path up their middles, wind up to an unknown place overhead where the heavy-footed eighth graders changing classes were called "the mules" by my first-grade teacher, and believing her, I pictured their sharp hooves on the linoleum.

My Grandfather Henry I remember in his sleeveless undershirt, home after a day's hard work on the farm at Fox Creek. His hide is tough and burnished wherever it has met the world—hands, face, forearms—but vulnerably white at the shoulders and throat. He is snapping his false teeth in and out of place, to provoke his grandchildren to hysterics.

As far as I know, no such snapshots exist in the authentic world. The citizens of my hometown ripped down the old school and quickly put to rest its picturesque decay. My grandfather always cemented his teeth in his head, and put on good clothes, before submitting himself to photography. Who wouldn't? When a camera takes aim at my daughter, I reach out and scrape the peanut butter off her chin. "I can't help it," I tell her, "it's one of those mother things." It's more than that. It's human, to want the world to see us as we think we ought to be seen.

You can fool history sometimes, but you can't fool the memory of your intimates. And thank heavens, because in the broad valley between real life and propriety whole herds of important truths can steal away into the underbrush. I hold that valley to be my home territory as a writer. Little girls wear food on their chins, school days are lit by ghostlight, and respectable men wear their undershirts at home. Sometimes there are fits of laughter and sometimes there is despair, and neither one looks a thing like its formal portrait.

For many, many years I wrote my stories furtively in spiral-bound notebooks, for no greater purpose than my own private salvation. But on April 1, 1987, two earthquakes hit my psyche on the same day. First, I brought home my own newborn baby girl from the hospital. Then, a few hours later, I got a call from New York announcing that a large chunk of my writing—which I'd tentatively pronounced a novel—was going to be published. This was a spectacular April Fool's Day. My life has not, since, returned to normal.

For days I nursed my baby and basked in hormonal euphoria, musing occasionally: all this—and I'm a novelist, too! *That*, though, seemed a slim accomplishment compared with laboring twenty-four hours to render up the most beautiful new human the earth had yet seen. The book business seemed a terrestrial affair of ink and trees and I didn't give it much thought.

In time my head cleared, and I settled into panic. What had I done? The baby was premeditated, but the book I'd conceived recklessly, in a closet late at night, when the restlessness of my insomniac pregnancy drove me to compulsive verbal intercourse with my own soul. The pages that grew in a stack were somewhat incidental to the process. They contained my highest hopes and keenest pains, and I didn't think anyone but me would ever see them. I'd bundled the thing up and sent it off to New York in a mad fit of housekeeping, to be done with it. Now it was going to be laid smack out for my mother, my postal clerk, my high school English teacher, anybody in the world who was willing to plunk down $16.95 and walk away with it. To find oneself suddenly published is thrilling—that is a given. But how appalling it also felt I find hard to describe. Imagine singing at the top of your lungs in the shower as you always do, then one day turning off the water and throwing back the curtain to see there in your bathroom a crowd of people, rapt, with videotape. I wanted to throw a towel over my head.

There was nothing in the novel to incriminate my mother or

the postal clerk. I like my mother, plus her record is perfect. My postal clerk I couldn't vouch for; he has tattoos. But in any event I never put real people into my fiction—I can't see the slightest point of that, when I have the alternative of inventing utterly subservient slave-people, whose every detail of appearance and behavior I can bend to serve my theme and plot.

Even so, I worried that someone I loved would find in what I'd written a reason to despise me. In fact, I was sure of it. My fiction is not in any way about my life, regardless of what others might assume, but certainly it is set in the sort of places I know pretty well. The protagonist of my novel, titled *The Bean Trees*, launched her adventures from a place called "Pittman, Kentucky," which does resemble a town in Kentucky where I'm known to have grown up. I had written: "Pittman was twenty years behind the nation in practically every way you can think of except the rate of teenage pregnancies. . . . We were the last place in the country to get the dial system. Up until 1973 you just picked up the receiver and said, Marge, get me my Uncle Roscoe. The telephone office was on the third floor of the Courthouse, and the operator could see everything around Main Street square. She would tell you if his car was there or not."

I don't have an Uncle Roscoe. But if I *did* have one, the phone operator in my hometown, prior to the mid-seventies, could have spotted him from her second-floor office on Main Street square.

I cherish the oddball charm of that town. Time and again I find myself writing love letters to my rural origins. Growing up in small-town Kentucky taught me respect for the astounding resources people can drum up from their backyards, when they want to, to pull each other through. I tend to be at home with modesty, and suspicious of anything slick or new. But naturally, when I was growing up there, I yearned for the slick and the new. A lot of us did, I think. We craved shopping malls and a swimming pool. We wanted the world to know we had once won the title "All Kentucky City," even though with sixteen-hundred souls we no more constituted a "city" than New Jersey is a Garden State, and we advertised this glorious prevarication for years and years on one of the town's few billboards.

Homely charm is a relative matter. Now that I live in a western city where shopping malls and swimming pools congest the landscape like cedar blight, I think back fondly on my home-town. But the people who live there now might rather smile about the quaintness of a *smaller* town, like nearby Morning Glory or Barefoot. At any rate, they would not want to discover themselves in my novel. I can never go home again, as long as I live, I reasoned. Somehow this will be reckoned as betrayal. I've photographed my hometown in its undershirt.

During the year I awaited publication, I decided to calm down. There were other ways to think about this problem:

1. If people really didn't want to see themselves in my book, they wouldn't. They would think to themselves, "She is writing about Morning Glory, and those underdogs are from far-ther on down Scrubgrass Road."

2. There's no bookstore in my hometown. No one will know.

In November 1988, bookstoreless though it was, my home-town hosted a big event. Paper banners announced it, and stores closed in honor of it. A crowd assembled in the town's largest public space—the railroad depot. The line went out the door and away down the tracks. At the front of the line they were plunking down $16.95 for signed copies of a certain book.

My family was there. The county's elected officials were there. My first-grade teacher, Miss Louella, was there, exclaiming to one and all: "I taught her to write!"

My old schoolmates were there. The handsome boys who'd spurned me at every homecoming dance were there.

It's relevant and slightly vengeful to confess here that I was not a hit in school, socially speaking. I was a bookworm who never quite fit her clothes. I managed to look fine in my school pictures, but as usual the truth lay elsewhere. In sixth grade I hit my present height of five feet almost nine, struck it like a gong, in fact, leaving behind self-confidence and any genuine need of a training bra. Elderly relatives used the term "fill out" when they spoke of me, as though they held out some hope I might eventually have some market value, like an underfed calf, if the hay crop was good. In my classroom I came to dread a game called Cooties, wherein one boy would brush against my shoulder and then chase the others around, threatening to pass on my apparently communicable lack of charisma. The other main victim of this game was a girl named Sandra, whose family subscribed to an unusual religion that mandated a Victorian dress code. In retrospect I can't say exactly what Sandra and I had in common that made us outcasts, except for extreme shyness, flat chests, and families who had their eyes on horizons pretty far beyond the hills of Nicholas County. Mine were not Latter-day Saints, but we read Thoreau and Robert Burns at home, and had lived for a while in Africa. My parents did not flinch from relocating us to a village beyond the reach of electricity, running water, or modern medicine (also, to my delight, conventional schooling) when they had a chance to do useful work there. They thought it was shameful to ignore a fellow human in need, or to waste money on trendy, frivolous things; they did not, on the other hand, think it was shameful to wear perfectly good hand-me-down dresses to school in Nicholas County. Ephemeral idols exalted by my peers, such as Batman, the Beatles, and the Hula Hoop, were not an issue at our house. And even if it took no more than a faint pulse to pass the fifth grade, my parents expected me to set my own academic goals, and then exceed them.

Possibly my parents were trying to make sure I didn't get pregnant in the eighth grade, as some of my classmates would shortly begin to do. If so, their efforts were a whale of a success. In my first three years of high school, the number of times I got asked out on a date was zero. This is not an approximate number. I'd caught up to other girls in social skills by that time, so I knew how to pretend I was dumber than I was, and make my own clothes. But these things helped only marginally. Popularity remained a frustrating mystery to me.

Nowadays, some of my city-bred friends muse about moving to a small town for the sake of their children. What's missing from their romantic picture of Grover's Corners is the frightening impact of insulation upon a child who's not dead center in the mainstream. In a place such as my hometown, you file in and sit down to day one of kindergarten with the exact pool of boys who will be your potential dates for the prom. If you wet your pants a lot, your social life ten years later will be—as they say in government reports—impacted. I was sterling on bladder control, but somehow could never shake my sixth-grade stigma.

At age seventeen, I was free at last to hightail it for new social pastures, and you'd better believe I did. I attended summer classes at the University of Kentucky and landed a boyfriend before I knew what had hit me, or what on earth one did with the likes of such. When I went on to college in Indiana I was astonished to find a fresh set of peers who found me, by and large, likable and cootie-free.

I've never gotten over high school, to the extent that I'm still a little surprised that my friends want to hang out with me. But it made me what I am, for better and for worse. From living in a town that listened in on party lines, I learned both the price and value of community. And I gained things from my rocky school years: A fierce wish to look inside of people. An aptitude for listening. The habit of my own company. The companionship of keeping a diary, in which I gossiped, fantasized, and invented myself. From the vantage point of invisibility I explored the psychology of the underdog, the one who can't be what others desire but who might still learn to chart her own hopes. Her story was my private treasure; when I wrote *The Bean Trees* I called her Lou Ann. I knew for sure that my classmates, all of them cool as Camaros back then, would not relate to the dreadful insecurities of Lou Ann. But I liked her anyway.

And now, look. The boys who'd once fled howling from her cooties were lined up for my autograph. Football captains, cheerleaders, homecoming queens were all there. The athlete who'd inspired in me a near-fatal crush for three years, during which time he never looked in the vicinity of my person, was there. The great wits who gave me the names Kingfish and Queen Sliver were there.

I took liberties with history. I wrote long, florid inscriptions referring to our great friendship of days gone by. I wrote slowly. I made those guys wait in line *a long time.*

I can recall every sight, sound, minute of that day. Every open, generous face. The way the afternoon light fell through the windows onto the shoes of the people in line. In my inventory of mental snapshots these images hold the place most people reserve for the wedding album. I don't know whether other people get to have Great Life Moments like this, but I was lucky enough to realize I was having mine, right while it happened. My identity was turning backward on its own axis. Never before or since have I felt all at the same time so cherished, so aware of old anguish, and so ready to let go of the past. My past had let go of *me*, so I could be something new: Poet Laureate and Queen for a Day in hometown Kentucky. The people who'd watched me grow up were proud of me, and exuberant over an event that put our little dot on the map, particularly since it wasn't an airline disaster or a child falling down a well. They didn't appear to mind that my novel discussed small-town life frankly, without gloss.

In fact, most people showed unsurpassed creativity in finding themselves, literally, on the printed page. "That's my car isn't it?" they would ask. "My service station!" Nobody presented himself as my Uncle Roscoe, but if he had, I happily would have claimed him.

It's a curious risk, fiction. Some writers choose fantasy as an approach to truth, a way of burrowing under newsprint and formal portraits to find the despair that can stow away in a happy childhood, or the affluent grace of a grandfather in his undershirt. In the final accounting, a hundred different truths are likely to reside at any given address. The part of my soul that is driven to make stories is a fierce thing, like a ferret: long, sleek, incapable of sleep, it digs and bites through all I know of the world. Given that I cannot look away from the painful things, it seems better to invent allegory than to point a straight, bony finger like Scrooge's mute Ghost of Christmas Yet to

Come, declaring, "Here you will end, if you don't clean up your act." By inventing character and circumstance, I like to think I can be a kinder sort of ghost, saying, "I don't mean *you*, exactly, but just give it some thought, anyway."

Nice try, but nobody's really fooled. Because fiction works, if it does, only when we the readers believe every word of it. Grover's Corners is Our Town, and so is Cannery Row, and Lilliput, and Gotham City, and Winesburg, Ohio, and the dreadful metropolis of *1984*. We have all been as canny as Huck Finn, as fractious as Scarlett O'Hara, as fatally flawed as Captain Ahab and Anna Karenina. I, personally, am Jo March, and if her author Louisa May Alcott had a whole new life to live for the sole pursuit of talking me out of it, she could not. A pen may or may not be mightier than the sword, but it is brassier than the telephone. When the writer converses privately with her soul in the long dark night, a thousand neighbors are listening in on the party line, taking it personally.

Nevertheless, I came to decide, on my one big afternoon as Homecoming Queen, that I would go on taking the risk of writing books. Miss Louella and all those football players gave me the rash courage to think I might be forgiven again and again the sin of revelation. I love my hometown as I love the elemental stuff of my own teeth and bones, and that seems to have come through to my hometown, even if I didn't write it up in its Sunday best.

I used to ask my grandfather how he could pull fish out of a lake all afternoon, one after another, while my line and bobber lay dazed and inert. This was not my Grandfather Henry, but my other grandfather, whose face I connected in childhood with the one that appears on the flip side of a buffalo nickel. Without cracking that face an iota, he was prone to uttering the funniest things I've about ever heard. In response to my question regarding the fishing, he would answer gravely, "You have to hold your mouth right."

I think that is also the secret of writing: attitude. Hope, unyielding faith in the enterprise. If only I hold my mouth right, keep a clear fix on what I believe is true while I make up my stories, surely I will end up saying what I mean. Then, if I offend someone, it won't be an accidental casualty. More likely, it will be because we actually disagree. I can live with that. The memory of my buffalo-nickel grandfather advises me still, in lonely moments: "If you never stepped on anybody's toes, you never been for a *walk*."

I learned something else, that November day, that shook down all I thought I knew about my personal, insufferable, nobody's-blues-can-touch-mine isolation of high school. Before the book signing was over, more than one of my old schoolmates had sidled up and whispered: "That Lou Ann character, the insecure one? I know you based her on me."

HOW MR. DEWEY DECIMAL
SAVED MY LIFE

A librarian named Miss Truman Richey snatched me from the jaws of ruin, and it's too late now to thank her. I'm not the first person to notice that we rarely get around to thanking those who've helped us most. Salvation is such a heady thing the temptation is to dance gasping on the shore, shouting that we are alive, till our forgotten savior has long since gone under. Or else sit quietly, sideswiped and embarrassed, mumbling that we really did know pretty much how to swim. But now that I see the wreck that could have been, without Miss Richey, I'm of a fearsome mind to throw my arms around every living librarian who crosses my path, on behalf of the souls they never knew they saved.

I reached high school at the close of the sixties, in the Commonwealth of Kentucky, whose ranking on educational

spending was I think around fifty-first, after Mississippi and whatever was below Mississippi. Recently Kentucky has drastically
changed the way money is spent on its schools, but back then,
the wealth of the county decreed the wealth of the school, and
few coins fell far from the money trees that grew in Lexington.
Our county, out where the bluegrass begins to turn brown, was
just scraping by. Many a dedicated teacher served out earnest
missions in our halls, but it was hard to spin silk purses out of a
sow's ear budget. We didn't get anything fancy like Latin or
Calculus. Apart from English, the only two courses of study that
ran for four consecutive years, each one building upon the last,
were segregated: Home Ec for girls and Shop for boys. And so I
stand today, a woman who knows how to upholster, color-coor-
dinate a table setting, and plan a traditional wedding—valuable
skills I'm still waiting to put to good use in my life.

As far as I could see from the lofty vantage point of age sixteen, there was nothing required of me at Nicholas County High
that was going to keep me off the streets; unfortunately we had
no streets, either. We had lanes, roads, and rural free delivery
routes, six in number, I think. We had two stoplights, which were
set to burn green in all directions after 6 P.M., so as not, should
the event of traffic arise, to slow anybody up.

What we *didn't* have included almost anything respectable
teenagers might do in the way of entertainment. In fact, there
was one thing for teenagers to do to entertain themselves, and it
was done in the backs of Fords and Chevrolets. It wasn't uphol-
stering skills that were brought to bear on those backseats, either.
Though the wedding-planning skills did follow.

I found myself beginning a third year of high school in a
state of unrest, certain I already knew what there was to know,
academically speaking—all wised up and no place to go. Some of

my peers used the strategy of rationing out the Science and Math classes between periods of suspension or childbirth, stretching their schooling over the allotted four years, and I envied their broader vision. I had gone right ahead and used the classes up, like a reckless hiker gobbling up all the rations on day one of a long march. Now I faced years of Study Hall, with brief interludes of Home Ec III and IV as the bright spots. I was developing a lean and hungry outlook.

We did have a school library, and a librarian who was surely paid inadequately to do the work she did. Yet there she was, every afternoon, presiding over the study hall, and she noticed me. For reasons I can't fathom, she discerned potential. I expect she saw my future, or at least the one I craved so hard it must have materialized in the air above me, connected to my head by little cartoon bubbles. If that's the future she saw, it was riding down the road on the back of a motorcycle, wearing a black leather jacket with "Violators" (that was the name of our county's motorcycle gang, and I'm not kidding) stitched in a solemn arc across the back.

There is no way on earth I really would have ended up a Violator Girlfriend—I could only dream of such a thrilling fate. But I was set hard upon wrecking my reputation in the limited ways available to skinny, unsought-after girls. They consisted mainly of cutting up in class, pretending to be surly, and making up shocking, entirely untrue stories about my home life. I wonder now that my parents continued to feed me. I clawed like a cat in a gunnysack against the doom I feared: staying home to reupholster my mother's couch one hundred thousand weekends in a row, until some tolerant myopic farm boy came along to rescue me from sewing-machine slavery.

Miss Richey had something else in mind. She took me by

the arm in study hall one day and said, "Barbara, I'm going to teach you Dewey Decimal."

One more valuable skill in my life.

She launched me on the project of cataloging and shelving every one of the, probably, thousand books in the Nicholas County High School library. And since it beat Home Ec III by a mile, I spent my study-hall hours this way without audible complaint, so long as I could look plenty surly while I did it. Though it was hard to see the real point of organizing books nobody ever looked at. And since it was my God-given duty in those days to be frank as a plank, I said as much to Miss Richey.

She just smiled. She with her hidden agenda. And gradually, in the process of handling every book in the room, I made some discoveries. I found *Gone With the Wind*, which I suspected my mother felt was kind of trashy, and I found Edgar Allan Poe, who scared me witless. I found that the call number for books about snakes is 666. I found William Saroyan's *Human Comedy*, down there on the shelf between Human Anatomy and Human Physiology, where probably no one had touched it since 1943. But I read it, and it spoke to me. In spite of myself I imagined the life of an immigrant son who believed human kindness was a tangible and glorious thing. I began to think about words like *tangible* and *glorious*. I read on. After I'd read all the good ones, I went back and read Human Anatomy and Human Physiology and found that I liked those pretty well too.

It came to pass in two short years that the walls of my high school dropped down, and I caught the scent of a world. I started to dream up intoxicating lives for myself that I could not have conceived without the books. So I didn't end up on a motorcycle. I ended up roaring hell-for-leather down the backroads of transcendent, reeling sentences. A writer. Imagine that.

The most important thing about the books I read in my rebellion is that they were not what I expected. I can't say I had no previous experience with literature; I grew up in a house full of books. Also, I'd known my way around the town's small library since I was tall enough to reach the shelves (though the town librarian disliked children and censored us fiercely) and looked forward to the Bookmobile as hungrily as more urbane children listened for the ice cream truck. So dearly did my parents want their children to love books they made reading aloud the center of our family life, and when the TV broke they took about two decades to get around to fixing it.

It's well known, though, that when humans reach a certain age, they identify precisely what it is their parents want for them and bolt in the opposite direction like lemmings for the cliff. I had already explained to my classmates, in an effort to get dates, that I was raised by wolves, and I really had to move on from there. If I was going to find a path to adult reading, I had to do it my own way. I had to read things I imagined my parents didn't want me looking into. Trash, like *Gone With the Wind*. (I think, now, that my mother had no real problem with *Gone With the Wind*, but wisely didn't let on.)

Now that I am a parent myself, I'm sympathetic to the longing for some control over what children read, or watch, or do. Our protectiveness is a deeply loving and deeply misguided effort to keep our kids inside the bounds of what we know is safe and right. Sure, I want to train my child to goodness. But unless I can invoke amnesia to blot out my own past, I have to see it's impossible to keep her inside the world I came up in. That world rolls on, and you can't step in the same river twice. The things that prepared me for life are not the same things that will move my own child into adulthood.

What snapped me out of my surly adolescence and moved me on were books that let me live other people's lives. I got to visit the Dust Bowl and London and the Civil War and Rhodesia. The fact that Rhett Butler said "damn" was a snoozer to me—I hardly noticed the words that mothers worried about. I noticed words like *colour bar,* spelled "colour" the way Doris Lessing wrote it, and eventually I figured out it meant racism. It was the thing that had forced some of the kids in my county to go to a separate school—which wasn't even a school but a one-room CME church—and grow up without plumbing or the hope of owning a farm. When I picked up *Martha Quest,* a novel set in southern Africa, it jarred open a door that was right in front of me. I found I couldn't close it.

If there is danger in a book like *Martha Quest,* and the works of all other authors who've been banned at one time or another, the danger is generally that they will broaden our experience and blend us more deeply with our fellow humans. Sometimes this makes waves. It made some at my house. We had a few rocky years while I sorted out new information about the human comedy, the human tragedy, and the ways some people are held to the ground unfairly. I informed my parents that I had invented a new notion called justice. Eventually, I learned to tone down my act a little. Miraculously, there were no homicides in the meantime.

Now, with my adolescence behind me and my daughter's still ahead, I am nearly speechless with gratitude for the endurance and goodwill of librarians in an era that discourages reading in almost incomprehensible ways. We've created for ourselves a culture that undervalues education (compared with the rest of the industrialized world, to say the least), undervalues breadth of experience (compared with our potential), downright discourages critical thinking (judging from what the majority of us watch and read), and distrusts foreign ideas. "Un-American," from what I hear, is meant to be an insult.

Most alarming, to my mind, is that we the people tolerate censorship in school libraries for the most bizarre and frivolous of reasons. Art books that contain (horrors!) nude human beings, and *The Wizard of Oz* because it has witches in it. Not always, everywhere, but everywhere, always something. And censorship of certain ideas in some quarters is enough to sway curriculums at the national level. Sometimes profoundly. Find a publishing house that's brave enough to include a thorough discussion of the principles of evolution in a high school text. Good luck. And yet, just about all working botanists, zoologists, and ecologists will tell you that evolution is to their field what germ theory is to medicine. We expect our kids to salvage a damaged earth, but in deference to the religious beliefs of a handful, we allow an entire generation of future scientists to germinate and grow in a vacuum.

The parents who believe in Special Creation have every right to tell their children how the world was made all at once, of a piece, in the year 4,004 B.C. Heaven knows, I tell my daughter things about economic justice that are just about as far outside the mainstream of American dogma. But I don't expect her school to forgo teaching Western history or capitalist economics on my account. Likewise, it should be the job of Special Creationist parents to make their story convincing to their children, set against the school's bright scenery of dinosaur fossils and genetic puzzle-solving, the crystal clarity of Darwinian logic, the whole glorious science of an evolving world that tells its own creation story. It cannot be any teacher's duty to tiptoe around religion, hiding objects that might raise questions at home. Faith, by definition, is impervious to fact. A belief that can be changed by new information was probably a scientific one, not a religious one, and science derives its value from its openness to revision.

If there is a fatal notion on this earth, it's the notion that wider horizons will be fatal. Difficult, troublesome, scary—yes, all that. But the wounds, for a sturdy child, will not be mortal. When I read Doris Lessing at seventeen, I was shocked to wake up from my placid color-blind coma into the racially segregated town I called my home. I saw I had been a fatuous participant in a horrible thing. I bit my nails to the quick, cast nets of rage over all I loved for a time, and quaked to think of all I had—still have—to learn. But if I hadn't made that reckoning, I would have lived a smaller, meaner life.

The crossing is worth the storm. Ask my parents. Twenty years ago I expect they'd have said, "Here, take this child, we will trade her to you for a sack of limas." But now they have a special shelf in their house for books that bear the family name on their spines. Slim rewards for a parent's thick volumes of patience, to be sure, but at least there are no motorcycles rusting in the carport.

My thanks to Doris Lessing and William Saroyan and Miss Truman Richey. And every other wise teacher who may ever save a surly soul like mine.

LIFE WITHOUT GO-GO BOOTS

Fashion nearly wrecked my life. I grew up beyond its pale, convinced that this would stunt me in some irreparable way. I don't think it has, but for a long time it was touch and go.

We lived in the country, in the middle of an alfalfa field; we had no immediate access to Bobbie Brooks sweaters. I went to school in the hand-me-downs of a cousin three years older. She had excellent fashion sense, but during the three-year lag her every sleek outfit turned to a pumpkin. In fifth grade, when girls were wearing straight shifts with buttons down the front, I wore pastel shirtwaists with cap sleeves and a multitude of built-in petticoats. My black lace-up oxfords, which my parents perceived to have orthopedic value, carried their own weight in the spectacle. I suspected people noticed, and I knew it for sure on the day

Billy Stamps announced to the lunch line: "Make way for the Bride of Frankenstein."

I suffered quietly, casting an ever-hopeful eye on my eighth-grade cousin whose button-front shifts someday would be mine. But by the time I was an eighth grader, everyone with an iota of social position wore polka-dot shirts and miniskirts. For Christmas, I begged for go-go boots. The rest of my life would be endurable if I had a pair of those white, calf-high confections with the little black heels. My mother, though always inscrutable near Christmas, seemed sympathetic; there was hope. Never mind that those little black heels are like skate blades in inclement weather. I would walk on air.

On Christmas morning I received white rubber boots with treads like a pair of Michelins. My mother loved me, but had missed the point.

In high school I took matters into my own hands. I learned to sew. I contrived to make an apple-green polyester jumpsuit that was supremely fashionable for about two months. Since it took me forty days and forty nights to make the thing, my moment of glory was brief. I learned what my mother had been trying to tell me all along: high fashion has the shelf life of potato salad. And when past its prime, it is similarly deadly.

Once I left home and went to college I was on my own, fashion-wise, having bypassed my cousin in stature and capped the arrangement off by moving to another state. But I found I still had to reckon with life's limited choices. After classes I worked variously as a house cleaner, typesetter, and artists' model. I could spend my wages on trendy apparel (which would be useless to me in any of my jobs, particularly the latter), or on the lesser gratifications of food and textbooks. It was a tough call, but I opted for education. This was Indiana and it was cold; when it wasn't cold, it was rainy. I bought an army surplus overcoat, with zip-out lining, that reached my ankles, and I found in my parents' attic a green pith helmet. I became a known figure on campus. Fortunately, this was the era in which army boots were a fashion option for coeds. And besides, who knew? Maybe under all that all-weather olive drab was a Bobbie Brooks sweater. My social life picked right up.

As an adult, I made two hugely fortuitous choices in the women's-wear department: first, I moved out West, where the buffalo roam and hardly anyone is ever arrested for being unstylish. Second, I became a novelist. Artists (also mathematicians and geniuses) are greatly indulged by society when it comes to matters of grooming. If we happen to look like an unmade bed, it's presumed we're preoccupied with plot devices or unifying theories or things of that ilk.

Even so, when I was invited to attend an important author event on the East Coast, a friend took me in hand.

"Writers are *supposed* to be eccentric," I wailed.

My friend, one of the people who loves me best in the world, replied: "Barbara, you're not eccentric, you're an anachronism," and marched me down to an exclusive clothing shop.

It was a very small store; I nearly hyperventilated. "You could liquidate the stock here and feed an African nation for a year," I whispered. But under pressure I bought a suit, and wore it to the important author function. For three hours of my life I was precisely in vogue.

Since then it has reigned over my closet from its dry-cleaner bag, feeling unhappy and out of place, I am sure, a silk ambassador assigned to a flannel republic. Even if I go to a chichi restaurant, the suit stays home. I'm always afraid I'll spill something on it; I'd be too nervous to enjoy myself. It turns out I would rather converse than make a statement.

Now, there is fashion, and there is *style*. The latter, I've found, will serve, and costs less. Style is mostly a matter of acting as if you know very well what you look like, thanks, and are just delighted about it. It also requires consistency. A friend of mine wears buckskin moccasins every day of her life. She has daytime and evening moccasins. This works fine in Arizona, but when my friend fell in love with a Tasmanian geologist and prepared to move to a rain forest, I worried. Moccasins instantaneously decompose in wet weather. But I should have known, my friend has sense. She bought clear plastic galoshes to button over her moccasins, and writes me that she's happy.

I favor cowboy boots. I don't do high heels, because you never know when you might actually have to get somewhere, and most other entries in the ladies-shoes category look to me like Ol' Dixie and Ol' Dobbin trying to sneak into the Derby, trailing their plow. Cowboy boots aren't trying. They say, "I'm no pump, and furthermore, so what?" That characterizes my whole uniform, in fact: oversized flannel shirts, jeans or cotton leggings, and cowboy boots when weather permits. In summer I lean toward dresses that make contact with the body (if at all) only on the shiatsu acupressure points; maybe also a Panama hat; and sneakers. I am happy.

I'm also a parent, which of course calls into question every decision one ever believes one has made for the last time. Can I raise my daughter as a raiment renegade? At present she couldn't care less. Maybe obsessions skip a generation. She was blessed with two older cousins whose sturdy hand-me-downs she has worn from birth, with relish. If she wasn't entirely a fashion plate, she also escaped being typecast. For her first two years she had no appreciable hair, to which parents can clamp those plastic barrettes that are gender dead giveaways. So when I took her to the park in cousin Ashley's dresses, strangers commented on her blue eyes and lovely complexion; when she wore Andrew's playsuits emblazoned with trucks and airplanes (why is it we only decorate our boys with modes of transportation?), people always commented on how strong and alert my child was—and what's his name?

This interests me. I also know it can't last. She's in school now, and I'm very quickly remembering what school is about: two parts ABCs to fifty parts Where Do I Stand in the Great Pecking Order of Humankind? She still rejects stereotypes, with extraordinary good humor. She has a dress-up collection to die for, gleaned from Goodwill and her grandparents' world travels, and likely as not will show up to dinner wearing harem pants, bunny ears, a glitter-bra over her T-shirt, wooden shoes, and a fez. But underneath it all, she's only human. I have a feeling the day might come when my daughter will beg to be a slave of conventional fashion.

I'm inclined to resist, if it happens. To press on her the larger truths I finally absorbed from my own wise parents: that she can find her own path. That she will be more valued for inward individuality than outward conformity. That a world plagued by poverty can ill afford the planned obsolescence of *haute couture*.

But a small corner of my heart still harbors the Bride of Frankenstein, eleven years of age, haunting me in her brogues and petticoats. Always and forever, the ghosts of past anguish compel us to live through our children. If my daughter ever asks for the nineties equivalent of go-go boots, I'll cave in.

Maybe I'll also buy her some of those clear plastic galoshes to button over them on inclement days.

THE HOUSEHOLD ZEN

In Barbara Pym's novel *Excellent Women*, published in 1952, there's a moment when our heroine pays a call on her new downstairs neighbor, a dubious kind of woman who wears trousers and is always dashing off to meetings of the Anthropological Society. When this woman answers the door, she shrugs without remorse at her unkempt apartment and declares, "I'm such a slut."

Wonderful word. Like so many others—gay, pill, roommate—it's acquired a sexual edge since the fifties, and it's too bad about *slut*, because the language needs a word to describe this particular relationship to housework. Something to tell the UPS man.

A select group of friends and I have formed a secret slut society. We wear trousers, we have fascinating work, and it's possible that the dust bunnies under our beds could be breeding dust bison. It's pretty shocking. In our lives we've seen revolutions in birth control and microchips and air bags, and all these are nothing compared to what's happened to housework. Interestingly, technology has nothing to do with it.

"We *had* a dishwasher, but my mother insisted you had to scrub and rinse every dish before you loaded it in," one of my colleagues in sluthood recalled as we sat around drinking something instant. The rest of us knew the story. The Kitchen Mystique. Dad gives Mom a microwave—a putative labor-saving device—and she reorganizes the whole kitchen as if the family had gone kosher, creating a supernatural order of kitchenware, some of which can go in the microwave and some of which will, she is sure, blow up in there. Melmac bombs.

My friend Jane dates a turning point in her life to the day in childhood when she *drew a diagram* of the vacuum cleaner before taking it out of the closet. When she finished vacuuming she put it away, every loop and coil scientifically in place, then silently watched her mother take it out and do it over again, claiming as always that it wasn't properly put away.

Cleaning houses in 1960 took ninety hours a week and the mind of a rocket scientist. Cleaning my house, in the nineties, takes a lick and a promise. Maybe fifteen adult-hours per week, for everything: laundry, dishes, a semiannual dust-bison roundup. The vacuum cleaner can stand on its head in the closet for all I care. I've discovered that almost any two things can be laundered together, and that the dishwasher will actually wash dishes if left to its own devices. Once in a blue moon my daughter's ballet tights will shrink and the forks won't entirely come clean; I donate them to the hand-me-down bag and run them through again, respectively. In these matters, it seems to me, an ounce of cure is worth a lifetime of prevention.

How did housework get to be so easy? I spent years wondering, until it dawned on me I was asking the wrong question. Why was it ever hard? I don't mean in the days of slogging clothes against rocks in the river, I mean in the days between our foremothers' competent Maytag and mine. Why did Donna Reed's house demand a full-time wife whereas mine asks for an occasional date? Because of historical necessity, pure and simple. In the fifties and sixties the economy boomed. One breadwinner could feed a family, and the social order demanded that Rosie the Riveter get out of the factory and into the kitchen. Betty Friedan, in *The Feminine Mystique*, chronicled the sociology of this period through an analysis of women's magazines, and it's pretty alarming to see how a culture gets rearranged by means of glossy paper. For the first time in history, through every means possible, housework was elevated from humble necessity to career status.

If the working-class women of my mother's generation had been born in any other time, they would have led other lives—not necessarily better or worse, but definitely other. A decade earlier, they might have built airplanes and let the devil and Hitler take the daily dusting. Ten years later, they could have had Ph.D.s in aeronautics. Women, unless they were quite wealthy, have always worked: in the house and out of the house, on the farm, in factories, sometimes caring for other people's kids, often leaving their own with the family herd under grandma's practiced eye. I've read that early in this century, when desperate families flooded into cities seeking work, leaving their rural support systems behind, female factory workers had to bundle their toddlers up on boards and hang them on hooks on the walls. At break time they'd unswaddle the kids and feed them. I like to mention this to anyone who suggests that modern day care is degrading the species.

Sometimes, when I'm trapped and have to listen to such stuff, I hear men of an evangelical bent explain that all our problems would end if women would just tend to housework and children as they have for two thousand years. Problem is, we are tending those things, but few of us have the option of doing it without also holding down a job that pays real money. Homemaking is moot if you're homeless. What the televangelists are invoking as the natural order is actually an artifice of a postwar economy, a kind of household that was practicable for just about twenty years. Not two thousand. Picture a medieval Donna Reed, if you will. Doesn't wash.

Whatever anyone might like to pretend, the fact is we're living in a country that can't—or in any event, doesn't—guarantee support for a spouse who does housework. And you don't get Toll House cookies without the toll. Behind the nostalgic call for women to return to tidying up the cottage is the supposition that some burly fellow will always be there to keep the wolf from the door. This fairy tale has lost its powers of persuasion. Half of all marriages undertaken since 1960 didn't last for the anticipated eternity. It's been a great disenchantment for all in the magic kingdom, no doubt, but the statistics on what follows are a shock that gets your feet back on the ground: after divorce, a man's expendable income is overwhelmingly likely to increase, while a woman's plummets, along with her children's standard of living. The reason for this is clear enough. Hours logged on *Kinder* and kitchen don't add up to tenure and a retirement plan.

Given that we have to make our own way in this big old world, it seems rude to try to make women (or men) feel guilty about neglecting the household operation. Cleanliness is next to godliness only if you're God's Wife. Guiding and nourishing a flock of the very young—your own or someone else's—is a career, there's no doubt about it. But housework is mostly about dirt. Other people's. The world's most renewable resource.

It seems incredible that some twenty years' worth of magazines could glorify the routine maintenance of marginally grateful sock-dropping families. I'm embarrassed about my own selfish participation in that experiment. For years I was determined to make it up to my mother. She has a job, now that we've all flown far from the nest, and I craved to ease her burden. Whenever she'd come to visit I would subtly try to demonstrate that housework could be the next best thing to nothing at all. "Let's all just pick up our plates," I'd say cheerfully at the end of a meal, "and *put them straight in the dishwasher.*" I felt shaky and heretical, as if I were suggesting to a priest that we spread Cheez Whiz on the wafers to make them more tasty. But I did it anyway. Right before Mom's very eyes I threw a pair of green socks in with the sheets.

Guess what? She already knew. She's no fool. Dad even cooks breakfast now.

I'm starting to see what my friend Jane, the scientific illustrator of vacuum cleaners, learned at age twelve: that in every profession, housewifery included, the necessity of feeling needed is the mother of inventive rules, some of which can't be penetrated by science. The process is separate from the product, and simple isn't what everybody needs, at every stage. Our work is how we define ourselves—so says every sociologist from Maria Montessori to Bruce Springsteen. If you work in the kitchen and have the mind of a rocket scientist, you're going to organize your cupboards like Mission Control. Nobody will know their way around it as well as you do. It needs to be that way. The gift of a microwave is an insult, if it suggests you could be replaced by the twist of a knob and a loud ding.

A generation of American women served their nation by being the Army of Moms, and they spent their creative force like the ancient Furies, whipping up cakes and handmade Christmas gifts and afterschool snacks, for a brief time in human history raising the art of homemaking high above the realm of dirt. Some of them fell as casualties to their era, and some won the medal of honor. Either way, they left a lot of us lucky baby boomers with strong teeth and bones and a warm taste of childhood in our mouths. No wonder those old boys are nostalgic. I am too.

Well, but I can also get nostalgic for the childhood of Laura Ingalls Wilder, until it dawns on me that not once, in any of those Little House books, does she discuss the real meaning of life without plumbing on howling cold prairie nights. Every epoch has its prizes and punishments, and there's no point in wishing my own were any different. The lot I drew in history was to belong to the generation of women groomed implicitly for wifehood, but who have ended up needing to win their bread rather than bake it. I've always been happy enough to do it, though now that I'm also supporting a child on my own, I occasionally wake up at night in a cold sweat on account of it; no part of my upbringing ever prepared me to hold this place at the head of the table. But it's a blessing, I think, to my girl, who is growing up convinced that women belong in the halls of discovery, production, and creation—messy enterprises all. It wouldn't even occur to her to doubt it. We've spent far more time together making kites and forts and scientifically mounted bug collections than working on hospital corners, and if her bed doesn't even get made, I'm the last to notice. Sluthood has its privileges, for children too.

Housework, like the Buddha, takes many forms, depending

on what is in your heart as you approach it. I personally am inclined to approach it the way governments treat dissent: ignore it until it revolts. If life were a different house of cards, though, and if housework were my life, you can bet it would acquire a heck of a lot of cachet. I would write book-length grocery lists, and serve meals that Proust would remember longer than those madeleines of his (whatever they were). Virtue in my living room would have the aroma of Lemon Fresh Pledge. My kitchen would be as cryptic as the streets of Venice, and I would be irreplaceable. The burly fellow in charge of keeping the wolf from my door would be lost without me, and I don't mean maybe. Some of those seemingly innocent dishes in my cabinet might be Molotov Melmac waiting to explode in the microwave. Really. I would never tell which ones.

SEMPER FI

Maybe this has happened to you: You are curled up on the sofa, with an afghan maybe, and the person you love is there too. You are female, because, I'm sorry, but I have the typewriter and you have to be what I say. And he is male. He is watching a contest of an athletic nature on TV, and you, well, you are present and accounted for. The contest is basketball, say, UCLA against Duke, in the NCAA playoffs. He's rooting for UCLA. You are confused. You were under the impression that he despised UCLA with a purple passion.

"Two weeks ago," you point out carefully, in the interest of scientific inquiry, "you were calling UCLA a bunch of galoots. You said they couldn't hit the side of a barn."

"Two weeks ago they were playing *us*. But now Arizona's out of the tournament," he explains, in that masculine sort of voice that can make any wild thing sound reasonable.

"But it's the same players," you persist, not wanting to make a fuss, but really. Once a galoot, always a galoot, it would seem to you. Nobody changes *that* much in two weeks, barring a religious experience, or steroids.

He sighs then, and patiently explains the hierarchy of loyalties: First you root for your home team. Then, if they're out of the picture, you root for other members of your conference.

"Even the *Sun Devils*?" you ask, dismayed. The Devils are your hometown team's nearest and bitterest rivals. The peak experience for a Devils fan is to sneak into Tucson and paint some important civic landmark such as the mayor in their school's colors.

"If the Devils were the only PAC 10 team left in the tournament, then sure, I'd want them to win." In an entirely even tone he says this perfectly preposterous thing, as if he is a chemistry professor announcing to an earnest, note-taking classroom that a new element in the periodic table of elements has been named after Donald Duck.

I've heard it many times. Not lately—it's been years, in fact, since Devils or Wildcats or Buffalo Bills or anything in tight pants and a helmet came into my house, because no one here is that interested. We tend to hold with Lorena Hickok, a columnist in the 1920s for the *Minneapolis Morning Tribune*, who observed of college football that "you might just as well put in your time watching a lot of ants running in and out of their hole. That is, if there isn't anything else you'd rather be doing right then." I'm sorry if I'm tipping sacred cows here. I don't mean to say I'm *above* watching organized sports. Possibly *below* it, for the fact is I'd rather watch ants. Draw your own conclusions.

But I am interested in sports as a concept, especially where it serves, like religion, as a touchstone for essential human longings. The entitlement to root for a different team each week is a little baffling, when held against other things we're supposed to take as self-evident. Love is eternal, isn't it? What is this slippery business, this hierarchy of shifting loyalties that glide in and out of place as methodically as the gears on a racing bike taking a hill? At first I suspected this creative fudge of allegiance had something to do with gender. I figured it was just one more of those mysteries withheld from women but revealed to men in their tender boyhoods, along with oil level vs. oil pressure, and how to believe you still look fine in a swimsuit once you've acquired love handles.

Determined to get to the bottom of it, I phoned a friend who has season tickets and wouldn't for love nor money miss an Arizona Wildcats game. And who is female. "Oh, no, I'd never root for the Sun Devils," she said without hesitation. "As far as I'm concerned it's Arizona or nobody."

Why? It's personal, she explained. After watching those six-foot-ten-and-still-growing boys play ball every Thursday night, you feel you know them. It's like they're your kids.

My friend paused; her tone was not all that maternal. "And let's face it," she added, "they've got great buns."

She allowed that her husband didn't share this outlook. "Oh, sure, he roots for other PAC 10 teams when they're not playing the Wildcats," she told me with a hint of scorn. "He'll root for anybody."

What is loyalty worth, if it's situational? This trend I was uncovering among certain sports fans reminded me of the song that suggests, if you can't be with your sweetie, you should love the sweetie who's handy. Whatever happened to "I'd rather be blue over you than happy with somebody new?"

Unquestionably, things like loyalty and territorial attachment *are* situational, from Candlestick Park to the Halls of Montezuma and in places far more ordinary. Even a dog, whose species has cornered the loyalty market, will show this weakness. I used to have one like that. She was a shepherd mongrel with a wild hair, half coyote. Her coyote instincts served her well for a good lifetime, steering her clear of what Darwin thought of as "nature red in tooth and claw" (though it was Tennyson who put it that way; Darwin couldn't have been that concise if his life depended on it). Out here in the desert, "tooth and claw" means prickly-pear spine and rattlesnake fang. My dog Jessie would often run congenially with small packs of coyotes, until she came within a stone's throw of the house. Then, brought up short against the sight of which side her bread was buttered on—which is to say, me—she would whirl around and make a big show of chasing her erstwhile friends out of the territory. I watched this happen dozens of times. One could argue conflicting genetic paradigms, or one could argue dog chow. Either way, Fido is an infidel.

It could be worse. Years ago as a graduate student I helped do a study of desert pupfish—a small, unglamorous species whose mating behavior is so opportunistic it would make Lolita blush. Pupfish live in ephemeral streams where populations fluctuate fairly drastically. When females are scarce, the male will hunt down a mate and swim faithfully by her side, for richer or for poorer, monogamous as Bob Cratchit. But when the tide turns and there's a surplus of females, the model-husband pupfish becomes a bantam-weight macho terror. He puffs out his little blue fins and claims a patch of river bottom as his private singles scene, performing all manner of gyrations to lure in the babes, who eventually do meander in to lay their eggs. Possibly they are rolling their eyes, muttering to one another about midlife crisis and the trophy wife. Darwin was right; nature is no picnic. It's an office party.

But it's not fair to cast this as a bad-boy business; females are no consistent models of fidelity either. Female elk are known to copulate with many males in the same day—and that's hardly the worst that can happen. The hills are alive with black-widow stories. A female praying mantis rewards her husband's conjugal exertions by eating his head; basically, that is their prenuptial agreement. And octopus mating, in its own special way, eclipses the tawdrily famous Bobbitts: the male octopus does not come equipped with a penis so he's obliged to offer his girlfriend a tough little packet of sperm (some valentine, that) by grasping it in a tentacle and shoving it down her breathing siphon. She responds to his overture by attempting to rip him apart. "These matings may be so violent," writes Robert A. Wallace in a forthright account, "that if the male has managed to insert his arm into the female's siphon, it may be literally torn from his body. After such an encounter, the female can be seen swimming alone, bearing the grisly memento of a previous coupling."

In a disenchanting revision of some cherished folklore, biologists are discovering that monogamy is rarer than unicorns in the animal world. Many species touted as mating for life—swans, bluebirds, Australian fairy wrens—are turning out to be hardcore sneaks. The tools of molecular genetics, similar to the tests used in human paternity suits, have shown that in the nest of the average fairy wren, one egg in five is sired by another wren's mate. Among all songbirds that have been examined in this way, the count is closer to one in three. It turns out the bluebird of happiness wrote the book on free love.

Our culture counts fidelity as a virtue, but where reproduction is concerned, it's more of a strategy. Monogamy is most likely to be practiced by creatures who have such pathetically helpless or numerous young it takes two frazzled parents to bring them to the self-supporting stage. Think of it as Darwinian family values: if a mate abandons the family, only to leave behind starved kids and nary a gene passed on, he or she is a biological dead end. So, for species in which the parenting demand is extreme, the biological directive that survives through the generations is the gene that sings out, "Be it ever so humble, there's no place like home."

This trend, predicting less promiscuity among species with a high parenting demand, bears out pretty well. Birds have the daunting egg chore and then a shrieking brood with mouths literally bigger than their bodies, so the parents do at least put up a show of keeping to the straight and narrow. Among mammals, monogamy is almost unknown, but it's adhered to by certain mice, voles, and pigs that have altricial (meaning pink, squirmy, frightfully helpless) young. Virtually all primates are promiscuous except the few species that always bear twins. In reptiles and amphibians, whose parenting style is best described as "hit and run," mate loyalty is out of the question.

Whether or not humans spent their millions of prelegal years being faithful to one true love is anyone's guess. If you look around now, you'll find every arrangement imaginable: wives who routinely take several husbands at once in the Himalayas; the reverse in Africa and Utah; serial monogamy in Polynesia and the contemporary United States. Many societies that aspire to monogamy are blunt about the loopholes, by recognizing a type of lineage anthropologists call avuncular: a child's paternal agent is the mother's brother, since he's the closest adult male who is known, with certainty, to be the child's blood relative.

We are a tough study. It's true our young are born fragile and witless as they come. But we are long-lived, too, and have so many opportunities to rethink the mate choice. "Rethink" is an important word—maybe we've traveled far enough from our origins so that biology has little to do with our amorous destiny. The wide variety of mating strategies we adopt across different cultures would suggest anything but biological determinism. But the battle of the sexes is such a persistent, bittersweet mystery the popular imagination seems convinced we are hard-wired for *la différence*. One extremely well-plowed argument goes this way: a male can increase his genes in the population by impregnating as many females as humanly possible, but it's to a female's advantage—since childbearing becomes her burden—to choose a mate who appears provident, loyal, thrifty, and inclined to stick around. So, the argument goes, men are predisposed to promiscuity, and women to being picky about their mates. Is it engraved upon us, this thing called adultery? It's an unanswerable question that seems to enthrall us no end. Physical anthropologists and sociobiologists have produced far more reams on the subject than Hugh Hefner ever did.

Sociobiology, which made a big splash in the seventies, threw some valuable light on the field of evolutionary biology, but it also threw some hooey into the kettle, where human behavior is concerned. Edward O. Wilson produced an incendiary book, *On Human Nature*, in which he asserted that there are biological bases for a large number (he implies, all) of the characteristics that are general enough to be called our "nature," and which we've integrated into our culture, political systems, and economy. I applaud Wilson (one of the world's preeminent biologists) for trying to bring humans back into the fold of nature. But he was roundly and rightly attacked, I think, for presuming that so much of human behavior—everything from armed combat to flirtation—

is directed by our genes. In seeking biological explanations Wilson provided almost no direct evidence for genetic control (as there is almost none to be found). He relied instead on analogy and "just-so" stories, suggesting that if a behavior appears to increase our likelihood of survival in certain contexts it must be biologically programmed. He ignored other levels of pressure—the social, material, and economic contexts—that influence decision making in the enormously flexible human brain. *On Human Nature* tried to draw us out onto the ice-thin proposition of biology as a new code of ethics: We are what we are, not because "God made us that way," but because four million years of natural selection did. And we'd better pay attention, Wilson warned, citing as a cautionary tale an example of enforced gender equality in an Israeli kibbutz, against which women rebelled and demanded more time with their children. (He neglected to mention that in this great experiment women were encouraged to value and perform men's work, but not the reverse, so women ended up doing both.) If we wish to change society, he wrote, "we can teach and reward and coerce. But in so doing, we must consider the price . . . measured in time and energy required for training and enforcement and in the less tangible currency of human happiness that must be spent to circumvent our innate predispositions." As science-based ethics replace those of religion, Wilson argued, our unconscious motives will drop out, we'll know what we're really capable of, and the truth will set us free.

Oh, but Dr. Wilson, which truth? Never in the deep blue sea will we ever be that conscious of our motives. The problem with identifying the biological roots of such things as sexism, aggression, and racism is that we're looking at our past through spectacles tinted with sexism, aggression, and racism. *On Human Nature* devotes a full chapter to the innateness of gender roles, in which women are passive and men naturally aggressive. (Not because God made us that way, but allegedly because it helped us survive.) Wilson began developing this line of thinking in an earlier book, *Sociobiology*, in which he wrote, "The populace of an American industrial city, no less than a band of hunter-gatherers in the Australian desert, is organized around [the nuclear family]. . . . During the day the women and children remain in the residential area while the men forage for game or its symbolic equivalent." He took this to be self-evident, and worked backward to construct a biological rationale for the arrangement. Stunningly, he did this in spite of the fact that in 1975, the year of the book's publication, 47 percent of all U.S. women aged sixteen and over were out working for the "symbolic equivalent," holding down two out of every five jobs.

In a similar feat of circular thinking, paleoanthropologists of the sixties and seventies presumed that human evolution was greatly influenced by pair bonding and a division of labor by gender. Evidence for this was the fact that skeletal remains of early hominids showed a marked size difference between males (larger) and females (smaller). These remains consisted only of fragments, never whole skeletons. They had been sorted into male and female with difficulty, frequently on the basis of nothing but their size!

Not only the answers we find but the very questions we ask, as scientists, are bathed in unconscious motives. *The Mismeasure of Man*, by Stephen Jay Gould, exposes the myriad ways science has been used throughout history to prove the superiority of Caucasian males, which happens to match the description of the scientists who did the work in question. Their work was earnest, painstaking, and dazzlingly blind to its own biases. European science of the eighteenth and nineteenth centuries presumed intel-

BARBARA KINGSOLVE

HIGH TIDE IN TUCSO

ESSAYS FROM
NOW OR NEVER

HarperCollins*Publishers*

ligence and human worthiness were things contained in the
head, and set itself like a dog on a bone to measuring skulls.
Thousands upon thousands of skulls: male and female, Cauca-sian
and Asian, Hottentot and Huron. Skulls of professionals were pit-
ted against those of clerks and laborers. The results are a testa-
ment to science's deep roots in creative interpretation and selective
oversight: the expected winners *always* came out on top. The
measurements were unconsciously bent, again and again, to make
it so. Just as Dr. Wilson forgot to notice that half of all modern
female "gatherers" were out "hunting," a nineteenth-century
physician named Samuel George Morton conducted his world-
renowned work on the essential character of human races without encumbering himself with contradictory evidence. Of the
Greenland Eskimo he wrote, "Their mental faculties, from
infancy to old age, present a continued childhood," and of the
Chinese, "So versatile are their feelings and actions, that they have
been compared to the monkey race, whose attention is perpetually changing from one object to another." Armed with these
foregone conclusions, he compared the brain sizes of these and
other races in his enormous skull collection with those of
Caucasians, by filling the cranial cavities with mustard seed or
lead BB shot, then pouring it back into a graduated cylinder and
reading the volume. Steven Jay Gould has carefully examined
Morton's data, wondering how such a respected scientist arrived
at a clear ranking of skull sizes (Caucasians largest, followed by
so-called Mongolians and American Indians, then Africans) that
cannot now be found to exist. Gould believes Morton didn't
consciously fudge his data, but did it in dozens of unconscious
ways, from selectively firing his assistants to failing to notice
the relationship between skull size and body size in all races, and
ultimately, by looking hard for what he needed to see. "Plausible

scenarios are easy to construct," Gould writes. "Morton, measuring by seed, picks up a threateningly large [African] skull, fills it lightly and gives it a few desultory shakes. Next, he takes a distressingly small Caucasian skull, shakes hard, and pushes mightily at the foramen magnum with his thumb."

It's nearly a vignette of black comedy to imagine Dr. Morton hunched over his skulls, with a racket of BB's rolling over his worktable and mustard seed crunching under his feet, as he labors to make the numbers of his science match the equations fixed in his heart. But in truth it's a horrific moment of history, for the data from his somber skullduggery were used to justify generations of genocide and slavery.

In this century, the biological determinists have laid down skulls and taken up testing. In 1912, when racism in America swelled on a rising tide of immigration, the U.S. Public Health Service hired psychologist H. H. Goddard to help screen out the imagined menace of inferior minds that were poised to contaminate the (equally imagined) pellucid American gene pool. Goddard, who invented the term "moron," created his own test for mental deficiency. Gould's *Mismeasure of Man* gives a remarkable account of how Goddard's test questions were fired at immigrants as they stepped bewildered and exhausted off the boat at Ellis Island. (Many had never before held a pencil, and had no possible frame of reference for understanding what was being asked of them.) Goddard arrived at these staggering results: 83 percent of Jews, 87 percent of Russians, 80 percent of Hungarians, and 79 percent of Italians were diagnosed as morons. At the time, no one paused to wonder how any nation could be carrying out its sundry business with four-fifths of its citizens punching in below the mental age of twelve; ethnic quotas on U.S. immigration were in place within the decade.

The explosive publication in 1994 of a book called *The Bell Curve* was an attempt to prove, yet again, the intellectual superiority of Caucasians. Written just in time to catch a new current of racism, the book drips with statistics and academic language, but its emotional heart seems bent on justifying the subordinate status of people of color in the U.S. Authors Richard Herrnstein and Charles Murray have made extravagant claims linking IQ and race, frequently basing them on the work of researchers who received grants from the Pioneer Fund, a bluntly pro-white organization that has long denounced school desegregation and advocated sterilization to eliminate so-called genetically unfit individuals or races from society. It's a case of science with the fingerprints of "motive" all over it.

No matter that *The Bell Curve* is made of very tired stuff, not much different from mismeasuring skulls; the book was taken seriously enough to capture the cover of *Newsweek*, tie up the headlines and news roundups, and sell like a house afire. The authors proved one thing, without a doubt: the privileged have not yet tired of hearing how righteously they came by their place at the table.

This whole line of inquiry, in which science is invoked to explain how we got where we are, is fatally tainted by what Hume called the "is-ought" problem. This is the philosophical error of confusing what *exists* with its *right* to exist, even though substance and scruples—what *is* and what *ought*—are entirely separate items, apples and oranges. When a mother says, "Boys will be boys," is she explaining her son's misbehavior, or predicting it, or forgiving it? Similarly, if we're allowed to talk on and on about white students scoring high marks on the Stanford-Binet, it's easy to slip a logical cog into who *ought* to get the better salary. And then there is all the sociobiological lore about male humans in the primordial social scene carrying on with big sticks, sniffing every wind for a shot at infidelity; after a while, it begins to suggest absolution, a certain now-and-forever slant on masculinity.

There's a simpler way to sum up the "is-ought" problem. A man I know, whenever he hears a story about philandering husbands or conniving wives, pipes up: "Sociobiology could explain it!"

And I intone: "Explain, maybe, but not excuse."

A creature with a big enough head to make a contract should have the sense to make one it can keep.

Of course, "See you later!" does not mean the same thing as "See you in court!" We make contracts on dozens of different levels. Hierarchies of urgency are understood, and so are the intrinsic values of different loyalties. Athletic contests are not marriages, although both are sweated out within bell jars of arbitrary rules. The advantage of sports is that they are quick—you decide what's important, stake your claim, and win or lose, you still go home unscathed. It might not look possible to an anthropologist from Venus, but life here really does last beyond the Super Bowl.

In a book called *The Stronger Women Get, the More Men Love Football*, Mariah Burton Nelson points out that sports are also about distinction, and perhaps that is why they assume such importance in our culture. "Who is better?" she writes. "One inch, one point, or one-hundredth of a second can differentiate winner from loser." Nelson lists at least six sports in which women and men now compete together at the elite level (dog-sled racing, horse racing, marathon swimming, equestrian events, rifle shooting, and auto racing), and many more recreational sports in which a wife and husband can typically find themselves evenly matched. And yet, she says, many people continue to rely hard on five games that showcase upper-body strength (football, baseball, basketball, boxing, and hockey) as reassurance of a certain order, gender-wise, in the universe.

Me, I bear in mind that women live seven years longer than men, on average, and figure that's the sport I'll sign up for.

So pick the rules that suit you, but just remember a game is no more than the sum of its parts: a stick, a ball, half an inch, two hundredths of a second. A cubic millimeter of muscle, or skull. A point of IQ. Come to think of it, things not much bigger than ants running into their hole.

All right, then. Back in your den, the game is winding down. Here is what you do: remind yourself that what you've been watching is a rigged arena. It's vastly popular simply because people flopped supine on furniture get to be muscular and sweaty by proxy and, for a short time, contrive their own rules about *what* makes *who* the *best*. Every day will dawn on a different "best," so the proxy contestants get to hitch their wagon to a new set of stars each time around. This says worlds about human nature, and nothing about real life. Game over, the river flows downhill again, and all the blue-eared pupfish go home to their mates.

You can give him a test, to make sure. "If I weren't around," you ask casually, "would you go out with my cousin Gloria? We're related—members of the same conference, you might say."

Your cousin Gloria is a blue-eyed version of Sonia Braga. Your sweetheart, though, is no fool. He gives you a hug and answers, "Don't be ridiculous. She's bowlegged."

Those are the rules. So what if there is no joy in Mudville, if at your house there's a place for everything, and every tentacle in its place.

THE MUSCLE MYSTIQUE

The baby-sitter surely thought I was having an affair. Years ago, for a period of three whole months, I would dash in to pick up my daughter after "work" with my cheeks flushed, my heart pounding, my hair damp from a quick shower. I'm loath to admit where I'd really been for that last hour of the afternoon. But it's time to come clean.

I joined a health club.

I went downtown and sweated with the masses. I rode a bike that goes nowhere at the rate of five hundred calories per hour. I even pumped a little iron. I can't deny the place was a lekking ground: guys stalking around the weight room like prairie chickens, nervously eying each other's pectorals. Over by the abdominal machines I heard some of the frankest pickup lines since eighth grade ("You've got real defined deltoids for a girl"). A truck perpetually parked out front had vanity plates that read: LFT WTS. Another one, PRSS 250, I didn't recognize as a vanity plate until I understood the prestige of bench pressing 250 pounds.

I personally couldn't bench press a fully loaded steam iron. I didn't join the health club to lose weight, or to meet the young Adonis who admired my (dubiously defined) deltoids. I am content with my lot in life, save for one irksome affliction: I am what's known in comic-book jargon as the ninety-eight-pound weakling. I finally tipped the scales into three digits my last year of high school, but "weakling" I've remained, pretty much since birth. In polite terminology I'm cerebral; the muscles between my ears are what I get by on. The last great body in my family was my Grandfather Henry. He wore muscle shirts in the days when they were known as BVDs, under his cotton work shirt, and his bronze tan stopped midbiceps. He got those biceps by hauling floor joists and hammering up roof beams every day of his life, including his last. How he would have guffawed to see a roomful of nearly naked bankers and attorneys, pale as plucked geese, heads down, eyes fixed on a horizon beyond the water cooler, pedaling like bats out of hell on bolted-down bicycles. I expect he'd offer us all a job. If we'd pay our thirty dollars a month to *him*, we could come out to the construction site and run up and down ladders bringing him nails. That's why I'm embarrassed about all this. I'm afraid I share his opinion of unproductive sweat.

Actually, he'd be more amazed than scornful. His idea of fun was watching Ed Sullivan or snoozing in a recliner, or ideally, both at once. Why work like a maniac on your day off? To keep your heart and lungs in shape. Of course. But I haven't noticed any vanity plates that say GD LNGS. The operative word here is vanity.

Standards of beauty in every era are things that advertise, usually falsely: "I'm rich and I don't have to work." How could you be a useful farmhand, or even an efficient clerk-typist, if you have long, painted fingernails? Four-inch high heels, like the bound feet of Chinese aristocrats, suggest you don't have to do *anything* efficiently, except maybe put up your tootsies on an ottoman and eat bonbons. (And I'll point out here that aristocratic *men* wore the first high heels.) In my grandmother's day, women of all classes lived in dread of getting a tan, since that betrayed a field worker's station in life. But now that the field hand's station is occupied by the office worker, a tan, I suppose, advertises that Florida and Maui are within your reach. Fat is another peculiar cultural flip-flop: in places where food is scarce, beauty is three inches of subcutaneous fat deep. But here and now, jobs are sedentary and calories are relatively cheap, while the luxury of time to work them off is very dear. It still gives me pause to see an ad for a weight-loss program that boldly enlists: "First ten pounds come off free!" But that is about the size of it, in this strange food-drenched land of ours. After those first ten, it gets expensive.

As a writer I could probably do my job fine with no deltoids at all, or biceps or triceps, so long as you left me those vermicelli-sized muscles that lift the fingers to the keyboard. (My vermicellis are *very* well defined.) So when I've writ my piece, off I should merrily go to build a body that says I don't really have a financial obligation to sit here in video-terminal bondage.

Well, yes. But to tell the truth, the leisure body and even the GD LNGS are not really what I was after when I signed up at Pecs-R-Us. What I craved, and long for still, is to be *strong*. I've never been strong. In childhood, team sports were my most reliable source of humiliation. I've been knocked breathless to the ground by softballs, basketballs, volleyballs, and once, during a wildly out-of-hand game of Red Rover, a sneaker. In every case I knew my teammates were counting on me for a volley or a double play or anyhow something more than clutching my stomach and rolling upon the grass. By the time I reached junior high I wasn't even the last one picked anymore. I'd slunk away long before they got to the bottom of the barrel.

Even now, the great mortification of my life is that visitors to my home sometimes screw the mustard and pickle jar lids back on so tightly *I can't get them open!* (The visitors probably think they are just closing them enough to keep the bugs out.) Sure, I can use a pipe wrench, but it's embarrassing. Once, my front gate stuck, and for several days I could only leave home by clambering furtively through the bougainvilleas and over the garden wall. When a young man knocked on my door to deliver flowers one sunny morning, I threw my arms around him. He thought that was pretty emotional, for florists' mums. He had no idea he'd just casually pushed open the Berlin Wall.

My inspiration down at the health club was a woman firefighter who could have knocked down my garden gate with a karate chop. I still dream about her triceps. But I've mostly gotten over my brief fit of muscle envy. Oh, I still make my ongoing, creative stabs at body building: I do "girl pushups," and some of the low-impact things from Jane Fonda's pregnant-lady workout book, even if I'm not. I love to run, because it always seems like there's a chance you might actually get somewhere, so I'll sometimes cover a familiar mile or so of our country road after I see my daughter onto the school bus. (The driver confessed that for weeks he thought I was chasing him; he never stopped.) And finally, my friends have given me an official item of exercise equipment that looks like a glob of blue putty, which you're supposed to squeeze a million times daily to improve your grip. That's my current program. The so-called noncompetitive atmosphere of the health club whipped me, hands down. Realistically, I've always known I was born to be a "before" picture. So I won't be seen driving around with plates that boast: PRSS 250.

Maybe: OPN JRS.

CIVIL DISOBEDIENCE AT BREAKFAST

I have a child who was born with the gift of focus, inclined to excel at whatever she earnestly pursues. Soon after her second birthday she turned to the earnest pursuit of languor, and shot straight through the ranks to world-class dawdler. I thought it might be my death.

Like any working stiff of a mother keeping the family presentable and solvent, I lived in a flat-out rush. My daughter lived on Zen time. These doctrines cannot find peace under one roof. I tried everything I could think of to bring her onto my schedule: five-minute countdowns, patient explanations of our itinerary, frantic appeals, authoritarianism, the threat of taking her to preschool *exactly* however she was dressed when the clock hit seven. (She went in PJs, oh delight! Smug as Brer Rabbit in the briars.) The more I tried to hurry us along, the more meticulously unhurried her movements became.

My brother pointed out that this is how members of the Japanese Parliament carry out a filibuster—by shuffling up to the voting box so extremely slowly it can take one person an hour to get across the room, and a month or two to get the whole vote in. It's called "cow walking," he reported. Perfect, I said. At my house we are having a Cow Life.

And that's how it was, as I sat at breakfast one morning watching my darling idle dangerously with her breakfast. I took a spectacularly deep breath and said, in a voice I imagined was calm, "We need to be going very soon. Please be careful not to spill your orange juice."

She looked me in the eye and coolly knocked over her glass.

Bang, my command was dead. Socks, shirt, and overalls would have to be changed, setting back the start of my workday another thirty minutes. Thirty-five, if I wanted to show her who was boss by enforcing a five-minute time-out. She knew exactly what she was doing. A filibuster.

I'd been warned the day would dawn when my sweet, tractable daughter would become a Terrible Two. And still this entirely predictable thing broadsided me, because in the beginning she was *mine*—as much a part of my body, literally, as my own arms and legs. The milk I drank knit her bones in place, and her hiccups jarred me awake at night. Children come to us as a dramatic coup of the body's fine inner will, and the process of sorting out "self" from "other" is so gradual as to be invisible to a mother's naked soul. In our hearts, we can't expect one of our own limbs to stand up one day and announce its own agenda. It's too much like a Stephen King novel.

Later in the day I called a friend to tell my breakfast war

story. She had a six-year-old, so I expected commiseration. The point of my call, really, was to hear that one could live through this and that it ended. Instead, my friend was quiet. "You know," she said finally, "Amanda never went through that. I worry about her. She works so hard to please everybody. I'm afraid she'll never know how to please herself."

A land mine exploded in the back of my conscience. My child was becoming all I'd ever wanted.

The way of a parent's love is a fool's progress, for sure. We lean and we lean on the cherished occupation of making ourselves obsolete. I applauded my child's first smile, and decoded her doubtful early noises to declare them "language." I touched the ground in awe of her first solo steps, as if she alone among primates had devised bipedal locomotion. Each of these events in its turn—more than triumph and less than miracle—was a lightening, feather by feather, of the cargo of anxious hope that was delivered to me with my baby at the slip of our beginning.

"We teach our children one thing only, as we were taught: to wake up," claims Annie Dillard. That's just about the whole truth, a parent's incantation. Wake up, keep breathing, look alive. It's only by forming separateness and volition that our children relieve us of the deepest parental dread: that they might somehow *not* wake up, after all, but fail to thrive and grow, remaining like Sleeping Beauty in the locked glass case of a wordless infancy. More times than I could count, in those early days, I was stopped in the grocery by some kindly matron who exclaimed over my burbling pastel lump of baby: "Don't you wish you could keep them like that forever?" Exactly that many times, I bit the urge to shout back, "Are you out of your mind?"

From the day she emerged open-mouthed in the world, I've answered my child's cries with my own gaping wonder, scrambling to part the curtains and show the way to wakefulness. I can think or feel no more irresistible impulse. In magnificent pantomime, I demonstrate to my small shadow the thousand and one ways to be a person, endowed with opinions. How could it be a surprise that after two years the lessons started to take? The shadow began to move of its own accord, exhibiting the skill of opinion by any means necessary. Barreling pell-mell through life was not my daughter's style; a mother ought to arrange mornings to allow time for communing with the oatmeal—that was her first opinion. How could I fail to celebrate this new red-letter day? There had been a time when I'd reduced my own personal code to a button on my blue-jeans jacket that advised: QUESTION AUTHORITY. A few decades later, the motto of my youth blazed resplendent on my breakfast table, the color of Florida sunshine. I could mop up, now, with maternal pride, or eat crow.

Oh, how slight the difference between "independent" and "ornery." A man who creates spectacular sculptures out of old car bodies might be a wonderful character, until he moves in next door. Children who lip off to their parents are cute in movies because they're in movies, and not in our life. Another of my brother's wise nuggets, offered over the phone one Saturday while I tried to manage family chaos and pour a cement porch foundation, was: "Remember, kids are better in the abstract than in the concrete." Of all kid abstractions, independence may be the hardest one to accept in the concrete, because we're told how we'll feel about it long before it arrives. It's the mother of all childhood stereotypes, the Terrible Twos.

Now there are stereotypes that encircle a problem like a darn good corral, and there are stereotypes that deliver a problem roaring to our doorstep, and I'm suspicious of this one, the Terrible Twos. If we'd all heard half so much about, say, the "Fat Fours," I'd bet dollars to donuts most four-year-olds would gain lots of weight, and those who didn't would be watched for the first sign of puffiness. Children are adept at becoming what we expect them to be. "Terrible" does not seem, by any stretch, to be a wise expectation. My Spanish-speaking friends—who, incidentally, have the most reliably child-friendly households in my acquaintance—tell me there's no translation for "Terrible Twos" in their language.

The global truth, I think, is that the twos are time-consuming and tidiness-impaired, but not, intrinsically, terrible. A cow in parliament is not a terrible cow. It's just a question of how it fits in with the plan.

The plan in our culture, born under the sign of freedom with mixed-message ascendant, is anyone's guess. The two developmental stages we parents are most instructed to dread—the twos and teens—both involve a child's formation of a sovereign identity. This, a plumb horror of assertive children, in the land of assertiveness training and weekend seminars on getting what you want through creative visualization. Expert advice on the subject of children's freedom is a pawnshop of clashing platitudes: We are to cultivate carefully the fragile stem of self-esteem. We are to consider a thing called "tough love," which combines militarist affection with house arrest, as remedy for adolescent misbehavior. We are to remember our children are only passing through us like precious arrows launched from heaven, but in most states we're criminally liable for whatever target they whack. The only subject more loaded with contradictions is the related matter of sex, which—in the world we've packaged for adolescents—is everywhere, visibly, the goal, and nowhere allowed. Let them eat it, drink it, wear it on their jeans, but don't for heaven's sakes pass out condoms, they might be inspired to *do* it. This is our inheritance, the mixed pedigree of the Puritans and Free Enterprise. We're to dream of our children growing up to be decision makers and trend setters, and we're to dream it through our teeth, muttering that a trend-setting toddler is a pain, and a teenager's decisions are a tour down the River Styx. How, then, to see it through?

The traditional camp says to hold the reins hard until the day we finally drop them, wish our big babies Godspeed, and send them out to run the world. I say, Good luck, it sounds like we'll have men and women with the mental experience of toddlers running domestic and foreign policy. (And, in fact, it sometimes appears that we do.) This is the parenting faction that also favors spanking. Studies of corporal punishment show, reliably, that kids who are spanked are more likely to be aggressive with their peers. For all the world, you'd think they were just little people, learning what they were taught.

I hold with those who favor allowing kids some freedom to work out problems their own way, and even make some messes, before we set them on Capitol Hill. I do not hold that this is easy. The most assiduous task of parenting is to divine the difference between boundaries and bondage. In every case, bondage is quicker. Boundaries, however carefully explained, can be reinterpreted creatively time and again. Yes, it's okay to pet the dog, and yes again on taking a bath, but *not* the dog *in* the tub. No to painting on the wall, no again to painting on the dog. I spent many years sounding to myself like Dr. Seuss: Not in a box! Not with a fox! Not on a train! Not in the rain!

The hardest boundaries to uphold are those that I know, in my heart, I have drawn for no higher purpose than my own convenience. I swore when I was pregnant I would never say to my child those stupid words "Because I said so!" Lord, have mercy. No contract I've ever signed has cost me so much. "Because I said so!" is not a real reason. But how about "Because if you do that again Mommy will scream, run into the bushes, pluck out the ovaries that made you, and cast them at the wild dogs." What price mental health? When your kid knocks over the orange juice, or ditches school, do you really have to listen to her inner wishes or can you just read the riot act?

Maybe both. Maybe there's not time for both right this minute—there never is, because life with children always bursts to fullness in the narrowest passages, like a life raft inflating in the emergency exit. If that's the case, then maybe the riot act now, and the other, listening to inner wishes, as soon as possible after you've worked free of the burning wreck.

During my short tenure as a parent I've relived my own childhood in a thousand ways while trying to find my path. Many of the things my parents did for me—most, I would say—are the things I want to do for my own child. Praise incessantly. Hold high expectations. Laugh, sing out loud, celebrate without cease the good luck of getting set down here on a lively earth.

But the world has changed since *Howdy Doody Time,* and some things nearly all parents did back then have been reconsidered. Spanking is one. Another, a little harder to define, has to do with structuring the family's time. My mother's job was me. But now I'm a mother with other work too, and fewer hours each day to devote to my main preoccupation of motherhood. I represent the norm for my generation, the throng of maternal employed, going about the honest work of the planet with gusto and generally no real alternative. The popular wisdom is that families used to be more kid-centered than they are now. I'm not so sure that's true. It's just different. My mother had kids to contend with from dawn till doom. She was (is) educated, creative, and much of the time the only people around for her to talk with had snakes in their pockets. My father worked very hard, as good fathers verily did. I had the guarantee of three squares daily, the run of several hundred acres of farms and wild Kentucky hills, the right to make a pet of anything nonvenomous, and a captive audience for theatrical projects. When my mother is canonized, I will testify that she really did sit through a hundred virtually identical productions, staged by my siblings and me, of the play titled approximately "The Dutch Boy Who Saved His Town by Putting His Finger in the Hole in the Dike." I have no idea why we did this. It seems truly obsessive. I can only offer as defense that we had a soft gray blanket with a hole in it, an irresistible prop. We took rave reviews for granted.

We also understood clearly that, during major family outings and vacations, our parents needed desperately to enjoy themselves. They bundled us into the back of the station wagon and begged us to go into hibernation for two thousand miles, so they could finish a conversation they'd started the previous autumn. I'm sure there were still plenty of times they sacrificed their vacation goals on the altar of my selfishness; I have forgotten these entirely. What I particularly remember instead is one nonstop auto trip to Key West, during which my sibs and I became bored beyond human limits. "Try counting to a million," my father suggested. And this is the point I am getting to: we actually did.

This seems amazing to me now. I could claim to be a victim, but that would be fatuous; my childhood was blessed. In the spectrum of the completely normal fifties family, nuclear units kept pretty much to themselves, and in the interest of everyone's survival, kids had to learn a decent show of obedience.

I'm amazed by the memory of counting to one million in a station wagon, not because I resent having done it myself, but because I can't imagine asking my daughter to do that, or, more to the point, *needing* for her to do it. When she and I head out on a car trip, we fall right into a fierce contest of White Horse Zit or license-plate alphabet. Childish enterprises, since they aren't my job, are in a sense my time off, my vacation. In spite of the well-publicized difficulties of balancing career and family, when I compare my life to my mother's I sometimes feel like Princess Grace. Each day I spend hours in luxurious silence, doing the work I most love; I have friends and colleagues who talk to me about interesting things, and never carry concealed reptiles. At the end of the day, when Camille and I are reunited after our daily cares, I'm ready for joyful mayhem.

For this reason I was also prepared to search through the pockets of my own soul on the day she and I arrived at our orange-juice impasse. I kept up a good authoritarian front at the time, but understood my daughter's implicit request. What was called for here was some Cow Time, stress free, no holds barred. I decided that after work we would go somewhere, out of the house, away from the call of things that require or provoke an orderly process. Together my two-year-old and I would waste the long last hours of an afternoon.

We went to the zoo. Not very far *into* the zoo, actually; we made it through the front gate and about twenty steps past, to the giant anteater den. There Camille became enraptured with a sturdy metal railing that was meant, I gather, to hold the public back from intimate contact with the giant anteaters. There was no danger, so I let her play on the metal bar.

And play.

After ten minutes I longed to pull her on toward the elephants, because frankly there's only so much looking a right-minded person can do at a giant anteater. But our agenda here was to have no agenda. I did my part. Looked again at those long anteating noses and those skinky anteating tongues.

Other children materialized on the bar. They clung and they dropped, they skinned the cat and impersonated tree sloths, until their parents eventually pulled them off toward the elephants. My eyes trailed wistfully after those departing families, but I knew I was being tested, and this time I knew I could win. I could refrain from asking my toddler to hurry up even longer than she could persist in sloth. After something less than an hour, she got down from the bar and asked to go home.

Five years have passed since then. Now it sometimes happens that Camille gets up, dresses herself in entirely color-coordinated clothes, and feeds the dog, all before the first peep of the alarm clock. I never cease to be amazed at this miracle, developmental biology. For any parent who needs to hear it today, I offer this: whatever it is, you can live through it, and it ends.

Plenty of psychologists have studied the effects of parents' behavior on the mental health of their children, but few have done the reverse. So Laurence Steinberg's study of 204 families with adolescents broke some new ground. All the families lived in Wisconsin but were otherwise diverse: rural, urban, white, black, brown, single-parented, remarried, nuclear. Steinberg uncovered a truth that crosses all lines: teenagers can make you crazy. Forty percent of the study parents showed a decline in psychological well-being during their children's adolescence. Steinberg even suggests that the so-called "midlife crisis" may be a response to living with teenagers, rather than to the onset of wrinkles and gray hair *per se*. The forty-four-year-old parent with a thirteen-year-old, it turns out, is far more disposed to crisis than the forty-four-year-old parent with an eight-year-old. Marital happiness tends to decline in households with teens, and single parents are more likely to experience difficulty with remarriage. But the study produced one hopeful note for the modern parent: in all family configurations, work is a buffer. Parents with satisfying careers had the best chance of sailing through the storms of their children's adolescence.

Here at last is a rallying cry for the throng of maternal employed. The best defense against a teenager's independence, and probably a toddler's as well, may simply be a matter of quitting before we're fired. Or not *quitting*, exactly, but backing off from eminent domain, happily and with dignity, by expressing ourselves in the serious pursuits and pleasures that we hold apart from parenting. Individuation goes both ways: we may feel less driven to shape a child in our own image if instead we can shape policy or sheet metal, or teach school, or boss around an employee or two. Luckiest of all is the novelist: I get to invent people who will live or die on the page, do exactly as I wish, *because I said so!*

I'm told it is terribly hard to balance career and family and, particularly, creativity. And it is, in fact. Good mothering can't be done by the clock. There are days I ache to throw deadlines to the wind and go hunt snipes. I wish for time to explain the sensible reason for every "no." To wallow in "yes," give over to a cow's timetable, stop the clock, stop watching the pot so it might splendidly boil.

I also long for more time of my own, and silence. My jaw

drops when I hear of the rituals some authors use to put themselves in the so-called mood to write: William Gass confesses to spending a couple of hours every morning photographing dilapidated corners of his city. Diane Ackerman begins each summer day "by choosing and arranging flowers for a Zenlike hour or so." She listens to music obsessively, then speed-walks for an hour, every single day. "I don't know whether this helps or not," she allows, in *A Natural History of the Senses*. "My muse is male, has the radiant, silvery complexion of the moon, and never speaks to me directly."

My muse wears a baseball cap, backward. The minute my daughter is on the school bus, he saunters up behind me with a bat slung over his shoulder and says oh so directly, "Okay, author lady, you've got six hours till that bus rolls back up the drive. You can sit down and write, *now*, or you can think about looking for a day job."

As a mother and a writer, I'd be sunk if either enterprise depended on corsages or magic. I start a good day by brushing my teeth; I don't know whether it helps or not, but it does fight plaque. I can relate at least to the utilitarian ritual of Colette, who began her day's writing after methodically picking fleas from her cat. The remarkable poet Lucille Clifton was asked, at a reading I attended, "Why are your poems always short?" Ms. Clifton replied, "I have six children, and a memory that can hold about twenty lines until the end of the day."

I would probably trade in my whole Great Books set for an epic-length poem from the pen of Lucille Clifton. But I couldn't wish away those six distracting children, even as a selfish reader, because I cherish Clifton's work precisely for its maternal passions and trenchant understanding of family. This is the fence we get to walk. I might envy the horses that prance unbridled across the pastures on either side of me, but I know if I stepped away from my fence into the field of "Only Work" or "Only Family," I would sink to my neck. I can hardly remember how I wrote before my child made a grown-up of me, nor can I think what sort of mother I would be if I didn't write. I hold with Dr. Steinberg: by working at something else I cherish, I can give my child room to be a chip off any old block she wants. She knows she isn't the whole of my world, and also that when I'm with her she's the designated center of my universe. On the day she walks away from my house for good, I'll cry and wave a hanky from my lonely balcony; then I'll walk to my study, jump for joy, and maybe do the best work of my life.

It's never easy to take the long view of things, especially in a society that conveys itself to us in four-second camera shots. But in a process as slow and complex as parenting, an eye to the future is an anchor. Raising children is a patient alchemy, which can turn applesauce into an athlete, ten thousand kissed bruises into one solid confidence, and maybe orneriness to independence. It all adds up. From the get-go I've been telling my child she is not just taking up space here, but truly valuable. If she's to believe it, I have to act as if I do. That means obedience is not an absolute value. Hurting people is out of the question, but an obsession with the anteater bar can and will be accommodated. I hope to hold this course as her obsessions grow more complex. For now, whenever the older, wiser parents warn, "Just wait till she's a teenager," I smile and say, "I'm looking forward to that." They think I am insane, impudent, or incredibly naïve. Probably I am. Call it creative visualization.

My time here is up today, for I'm being called to watch a theatrical production entitled approximately "The Princess Fairy Mermaids Who Save the Castle by Murderizing the Monsters and Then Making Them Come Back Alive with Fairy Dust and Be Nice." I've seen this show before. Some days I like it, especially when they tie up the monster with Day-Glo shoelaces and pantyhose. Other days my mind drifts off to that spare, uncluttered studio where I will arrange flowers, Zenlike, when I'm sixty. I'll write great things, and I'll know once and for all the difference between boundaries and bondage.

SOMEBODY'S BABY

As I walked out the street entrance to my newly rented apartment, a guy in maroon high-tops and a skateboard haircut approached, making kissing noises and saying, "Hi, gorgeous."

Three weeks earlier, I would have assessed the degree of malice and made ready to run or tell him to bug off, depending. But now, instead, I smiled, and so did my four-year-old daughter, because after dozens of similar encounters I understood he didn't mean me but *her*.

This was not the United States.

For most of the year my daughter was four we lived in Spain, in the warm southern province of the Canary Islands. I struggled with dinner at midnight and the subjunctive tense, but my only genuine culture shock reverberated from this earthquake of a fact: people there like kids. They don't just say so, they *do*. Widows in black, buttoned-down CEOs, purple-sneakered teenagers, the butcher, the baker, all would stop on the street to have little chats with my daughter. Routinely, taxi drivers leaned out the window to shout "*Hola, guapa!*" My daughter, who must have felt my conditioned flinch, would look up at me wide-eyed and explain patiently, "I *like* it that people think I'm pretty." With a mother's keen myopia I would tell you, absolutely, my daughter is beautiful enough to stop traffic. But in the city of Santa Cruz, I have to confess, so was every other person under the height of one meter. Not just those who conceded to be seen and not heard. Whenever Camille grew cranky in a restaurant (and really, what do you expect at midnight?) the waiters flirted and brought her little presents, and nearby diners looked on with that sweet, wistful gleam of eye that I'd thought diners reserved for the dessert tray. What I discovered in Spain was a culture that held children to be its meringues and éclairs. My own culture, it seemed to me in retrospect, tended to regard children as a sort of toxic-waste product: a necessary evil, maybe, but if it's not our own we don't want to see it or hear it or, God help us, smell it.

If you don't have children, you think I'm exaggerating. But if you've changed a diaper in the last decade you know exactly the toxic-waste glare I mean. In the U.S. I have been told in restaurants: "We come here to get *away* from kids." (This for no infraction on my daughter's part that I could discern, other than being visible.) On an airplane I heard a man tell a beleaguered woman whose infant was bawling (as I would, to clear my aching ears, if I couldn't manage chewing gum): "If you can't keep that thing quiet, you should keep it at home."

Air travel, like natural disasters, throws strangers together in unnaturally intimate circumstances. (Think how well you can get to know the bald spot of the guy reclining in front of you.)

Consequently airplanes can be a splendid cultural magnifying glass. On my family's voyage from New York to Madrid we weren't assigned seats together. I shamelessly begged my neighbor—a forty-something New Yorker traveling alone—if she would take my husband's aisle seat in another row so our air-weary and plainly miserable daughter could stretch out across her parents' laps. My fellow traveler snapped, "No, I have to have the window seat, just like you *had* to have that baby."

As simply as that, a child with needs (and ears) became an inconvenient *thing*, for which I was entirely to blame. The remark left me stunned and, as always happens when someone speaks rudely to me, momentarily guilty: yes, she must be right, conceiving this child was a rash, lunatic moment of selfishness, and now I had better be prepared to pay the price.

In the U.S.A., where it's said that anyone can grow up to be President, we parents are left pretty much on our own when it comes to the Presidents-in-training. Our social programs for children are the hands-down worst in the industrialized world, but apparently that is just what we want as a nation. It took a move to another country to make me realize how thoroughly I had accepted my nation's creed of every family for itself. Whenever my daughter crash-landed in the playground, I was startled at first to see a sanguine, Spanish-speaking stranger pick her up and dust her off. And if a shrieking bundle landed at *my* feet, I'd furtively look around for the next of kin. But I quickly came to see this detachment as perverse when applied to children, and am wondering how it ever caught on in the first place.

My grandfathers on both sides lived in households that were called upon, after tragedy struck close to home, to take in orphaned children and raise them without a thought. In an era of shortage, this was commonplace. But one generation later that kind of semipermeable household had vanished, at least among the white middle class. It's a horrifying thought, but predictable enough, that the worth of children in America is tied to their dollar value. Children used to be field hands, household help, even miners and factory workers—extensions of a family's productive potential and so, in a sense, the property of an extended family. But *precious* property, valued and coveted. Since the advent of child-labor laws, children have come to hold an increasingly negative position in the economy. They're spoken of as a responsibility, a legal liability, an encumbrance—or, if their unwed mothers are on welfare, a mistake that should not be rewarded. The political shuffle seems to be about making sure they cost us as little as possible, and that their own parents foot the bill. Virtually every program that benefits children in this country, from *Sesame Street* to free school lunches, has been cut back in the last decade—in many cases, cut to nothing. If it takes a village to raise a child, our kids are knocking on a lot of doors where nobody seems to be home.

Taking parental responsibility to extremes, some policymakers in the U.S. have seriously debated the possibility of requiring a license for parenting. I'm dismayed by the notion of licensing an individual adult to raise an individual child, because it implies parenting is a private enterprise, like selling liquor or driving a cab (though less lucrative). I'm also dismayed by what it suggests about innate fitness or nonfitness to rear children. Who would devise such a test? And how could it harbor anything but deep class biases? Like driving, parenting is a skill you learn by doing. You keep an eye out for oncoming disasters, and know when to stop and ask for directions. The skills you have going into it are hardly the point.

The first time I tried for my driver's license, I flunked. I was sixteen and rigid with panic. I rolled backward precariously while starting on a hill; I misidentified in writing the shape of a railroad crossing sign; as a final disqualifying indignity, my VW beetle—borrowed from my brother and apparently as appalled as I—went blind in the left blinker and mute in the horn. But nowadays, when it's time for a renewal, I breeze through the driver's test without thinking, usually on my way to some other errand. That test I failed twenty years ago was no prediction of my ultimate competence as a driver, anymore than my doll-care practices (I liked tying them to the back of my bike, by the hair) were predictive of my parenting skills (heavens be praised). Who really understands what it takes to raise kids? That is, until after the diaper changes, the sibling rivalries, the stitches, the tantrums, the first day of school, the overpriced-sneakers standoff, the first date, the safe-sex lecture, and the senior prom have all been negotiated and put away in the scrapbook?

While there are better and worse circumstances from which to launch offspring onto the planet, it's impossible to anticipate just who will fail. One of the most committed, creative parents I know plunged into her role through the trapdoor of teen pregnancy; she has made her son the center of her life, constructed a large impromptu family of reliable friends and neighbors, and absorbed knowledge like a plant taking sun. Conversely, some of the most strained, inattentive parents I know are well-heeled professionals, self-sufficient but chronically pressed for time. Life takes surprising turns. The one sure thing is that no parent, ever, has turned out to be perfectly wise and exhaustively provident, 1,440 minutes a day, for 18 years. It takes help. Children are not commodities but an incipient world. They thrive best when their upbringing is the collective joy and responsibility of families, neighborhoods, communities, and nations.

It's not hard to figure out what's good for kids, but amid the noise of an increasingly antichild political climate, it can be hard to remember just to go ahead and do it: for example, to vote to raise your school district's budget, even though you'll pay higher taxes. (If you're earning enough to pay taxes at all, I promise, the school needs those few bucks more than you do.) To support legislators who care more about afterschool programs, affordable health care, and libraries than about military budgets and the Dow Jones industrial average. To volunteer time and skills at your neighborhood school and also the school across town. To decide to notice, rather than ignore it, when a neighbor is losing it with her kids, and offer to baby-sit twice a week. This is not interference. Getting between a ball player and a ball is interference. The ball is inanimate.

Presuming children to be their parents' sole property and responsibility is, among other things, a handy way of declaring problem children to be someone else's problem, or fault, or failure. It's a dangerous remedy; it doesn't change the fact that somebody else's kids will ultimately be in your face demanding *now* with interest what they didn't get when they were smaller and had simpler needs. Maybe in-your-face means breaking and entering, or maybe it means a Savings and Loan scam. Children deprived—of love, money, attention, or moral guidance—grow up to have large and powerful needs.

Always there will be babies made in some quarters whose parents can't quite take care of them. Reproduction is the most invincible of all human goals; like every other species, we're only here because our ancestors spent millions of years refining their act as efficient, dedicated breeders. If we hope for only sane, thoughtful people to have children, we can wish while we're at it for an end to cavities and mildew. But unlike many other species we are social, insightful, and capable of anticipating our future. We can see, if we care to look, that the way we treat children—*all* of them, not just our own, and especially those in great need— defines the shape of the world we'll wake up in tomorrow. The most remarkable feature of human culture is its capacity to reach beyond the self and encompass the collective good.

It's an inspiring thought. But in mortal fact, here in the U.S. we are blazing a bold downhill path from the high ground of "human collective," toward the tight little den of "self." The last time we voted on a school-budget override in Tucson, the newspaper printed scores of letters from readers incensed by the very possibility: "I don't have kids," a typical letter writer declared, "so why should I have to pay to educate other people's offspring?" The budget increase was voted down, the school district progressed from deficient to desperate, and I longed to ask that miserly nonfather just *whose* offspring he expects to doctor the maladies of his old age.

If we intend to cleave like stubborn barnacles to our great American ethic of every nuclear family for itself, then each of us had better raise and educate offspring enough to give us each day, in our old age, our daily bread. If we don't wish to live by bread alone, we'll need not only a farmer and a cook in the family but also a home repair specialist, an auto mechanic, an accountant, an import-export broker, a forest ranger, a therapist, an engineer, a musician, a poet, a tailor, a doctor, and at least three shifts of nurses. If that seems impractical, then we can accept other people's kids into our lives, starting now.

It's not so difficult. Most of the rest of the world has got this in hand. Just about any country you can name spends a larger percentage of its assets on its kids than we do. Virtually all industrialized nations have better schools and child-care policies. And while the U.S. grabs headlines by saving the occasional baby with heroic medical experiments, world health reports (from UNESCO, USAID, and other sources) show that a great many other parts of the world have lower infant mortality rates than we do—not just the conspicuously prosperous nations like Japan and Germany, but others, like Greece, Cuba, Portugal, Slovenia—simply because they attend better to all their mothers and children. Cuba, running on a budget that would hardly keep New York City's lights on, has better immunization programs and a higher literacy rate. During the long, grim haul of a thirty-year economic blockade, during which the United States has managed to starve Cuba to a ghost of its hopes, that island's child-first priorities have never altered.

Here in the land of plenty a child dies from poverty every fifty-three minutes, and TV talk shows exhibit teenagers who pierce their flesh with safety pins and rip off their parents every way they know how. All these punks started out as somebody's baby. How on earth, we'd like to know, did they learn to be so isolated and selfish?

My second afternoon in Spain, standing in a crowded bus, as we ricocheted around a corner and my daughter reached starfishwise for stability, a man in a black beret stood up and gently helped her into his seat. In his weightless bearing I caught sight of the decades-old child, treasured by the manifold mothers of his neighborhood, growing up the way leavened dough rises surely to the kindness of bread.

I thought then of the woman on the airplane, who was obviously within her rights to put her own comfort first, but whose withheld generosity gave my daughter what amounted to a sleepless, kicking, squirming, miserable journey. As always happens two days after someone has spoken to me rudely, I knew exactly what I should have said: Be careful what you give children, for sooner or later you are sure to get it back.

PARADISE LOST

The Canary Islands weren't named for birds, but dogs. Pliny the Elder wrote of "Canaria, so called from the multitude of dogs [canis] of great size." In Pliny's day this archipelago, flung west from the coast of Morocco, was the most westerly place imaginable. All maps started here. For fourteen centuries Arabs, Portuguese, and eventually the Spanish came this far, and no farther; it remained Meridian Zero. When Columbus gathered the force to head west and enlarge the map, it was from La Gomera, in the Canaries, that he sailed.

I went to the Canaries for nearly a year, to find new stories to tell, and to grow comfortable thinking in Spanish. Or so I said; the truth is closer to the bone. It was 1991, and in the U.S. a clamor of war worship had sprung like a vitriolic genie from the riveted bottles we launched on Baghdad. Yellow ribbons swelled from suburban front doors, so puffy and ubiquitous as to seem folkloric. But this folklore, a prayer of godspeed to the killers, allowed no possibility that the vanquished might also be human. I grew hopeless, then voiceless. What words could I offer a place like this? Five hundred years after colonialism arrived in the New World, I booked a return passage.

Subtropical Europe seemed an idyllic combination of wild and tame: socialized health care and well-fed children, set in a peaceful tangle of banana trees and wild poinsettias. We settled in Tenerife's capital city, Santa Cruz, in a walk-up apartment that was tiny by U.S. standards, average by European, and anyhow what we could afford. I soon got used to living in a small space. The walls vibrated pleasantly with my upstairs neighbor's piano sonatas. I planted tomatoes and basil in pots on the balcony. My daughter became bilingual without realizing it, continuing to chatter in Spanish as I walked her home from kindergarten. In the afternoons she and I made forays to the bright, rowdy markets, to the beach, to wherever the green city buses would take us. We sat in sidewalk cafés on the harbor, watching cars go by. Behind the cars, enormous ships passed by on a lane of water not visible from our vantage point, so it looked as if ocean liners were sliding majestically up and down the Avenida de Anaga. In the park we collected round wooden jacaranda pods with toothy openings like small dragon mouths. We grew accustomed to the remarkable habit of walking there, perfectly safe, after dark. We did not miss the New World.

I set my writing desk against the apartment's front window, from which I could look down into the tops of the great fig trees that lined the street below—a broad boulevard named for General Franco, the distinguished despot and friend to Hitler. (My friends who sent me letters there will vouch for this, my astonishing fascist address.) So much for the innocence of this place, whose Spanish charm—like the whole world, apparently—is built on the bones of the vanquished. What new stories were here to tell? Instead of writing, I took to staring at the apartment across the street, also three floors up, where at night a fellow insomniac haunted his spartan balcony. I considered blinking my lights at him. I began to imagine a whole secret world of signals: A woman who sits on her balcony each morning drinking coffee, while the stranger across the street does the same. One day she buys a fern for her balcony, and the following day so does he. Then she buys a geranium, so does he. She fills her balcony, crowds it with flowers, so that he will too. Why? To watch him prove his devotion? Because she feels sorry for him, and wants him to drink his coffee in the lively embrace of a garden? Or simply because she has power over him? If that is the case, then she will take the plants away again, one by one, leaving him with nothing in the end. In my despondent state I could think of no happier ending. Power, like space, it seemed to me, would always get used. People expand and bloat to fill it.

My mind staggered and found nothing of use to tell. A small, squat spider patrolled the window casing above my writing desk. Its two white forelegs moved continually in a single repeated gesture: a scooping motion toward its mouth, like a mute beggar asking for bread. The spider became the muse of my empty page, asking, asking, asking to be filled.

In September Camille made plans to spend a weekend with a school friend. Given the chance to get away, briefly, I found I needed to go. I didn't know why, but I knew where. La Gomera.

Six of the seven Canary Islands have airports, the better to accommodate hasty visits from European sun worshipers. But any traveler who wants to approach the seventh, most secretive Canary—La Gomera—must take the sea road as Columbus did. I found myself that kind of traveler, in no particular hurry on a bright Saturday. I'd been told dolphins liked to gambol in the waves in this channel, and that sighting them brings good luck. I was ready for some luck. The sun on the pointed waves was hard as chipped flint, but I stared anyway, awaiting revelation.

The ferry from Tenerife to La Gomera churned away from a southern resort town with a bleached, unimaginative skyline of tourist hotels. For reasons difficult to fathom or appreciate, the brown hills dropping away behind the port displayed giant white letters spelling out HOLLYWOOD. An hour and a half ahead of us lay tiny La Gomera, where the hills don't yet speak English or anything else.

Among urban Canarians, La Gomera has a reputation for backwardness, and the Gomerans themselves are sometimes likened to Guanches—the tall, blue-eyed, goat-herding aboriginals whom the Spaniards found here and promptly extinguished in the fifteenth century. No one knows where they came from, though it's a good guess that they were related to the tall, blue-eyed Berbers who still roam the western Sahara. Throughout the Canaries, the Guanches herded goats, made simple red-clay pottery, and followed the lifestyle known as Neolithic, living out their days without the benefit of metal. They were farmers, not fishers; anthropologists insist these island people had no boats. On La Gomera they used a type of language unique in the world, which was not spoken but *whistled*. This exotic means of communication, called *silbo*, could traverse the great distances that routinely separate neighbors on an island cut through and through with steep, uncrossable gorges. (Whistling carries its subtleties over distance in a way that shouting can't.) I'd been told by many Canarians that the silbo has died out completely. But others claimed it still persists in some corners, along with pottery making and farming with the muscle of human and ox. I made a pact for the crossing: if I see dolphins in the channel, I'll believe the rest of the story.

The blue cliffsides of La Gomera seemed close enough to Tenerife to reach by means of a strong backstroke. It's hard to imagine living on islands this small, in plain view of other land, and never being stirred to build a boat. In fact, I know of an anthropologist who studied the archaeological record of the Guanches, but could not convince her colleagues that their culture shunned the sea. People with such mysterious motives seem more legendary than real. That's the great problem, I suppose, with becoming extinguished.

Just beyond the rushing ferry's shroud of spray, the dolphins appeared to me, slick and dark, rolling like finned inner tubes in the Atlantic.

San Sebastián de La Gomera is the port from which Columbus set sail for the New World. Elsewhere on earth, the approaching quincentennial anniversary of that voyage had been raising a lot of fuss, but here at the point of origin all was quiet. Fishing boats sat like sleeping gulls in the harbor, rolling in the ferry's wake. A store in the port sold T-shirts with the ambiguous message "Aquí partió Colón"—Columbus departed from here. So did everyone else, apparently. San Sebastián's narrow streets were empty save for long shadows of fig trees and a handful of noontime shoppers. I claimed my bantam-weight rental car and drove up a steep, cobbled hill to the *parador* overlooking the harbor.

The hotel, Parador Condé de La Gomera, is an old, elegant replica of an estate that stood here in Columbus's time. The massive front door leads to a cool interior of cut-stone archways and dark carved woodwork. Passages open to bright courtyards, where potted ferns grow head high and higher, brushing the door frames. The hallways turn out everywhere onto hidden sitting areas dappled with light, each one arranged like a perfect photograph. Easy enough a life, to stay forever in the paradise of San Sebastián. Columbus came close to doing it. Gomerans love to tell the story of how he delayed his first historic voyage for many months—nearly cashed it in altogether—having settled down here comfortably with the widow of the first Count of La Gomera, Beatriz de Bobadilla.

The balcony of my room overlooked the tops of palms and tamarinds leaning perilously over the edge of the cliff, and far below, the harbor. From a rocking chair on the balcony I watched the ferry that had brought me here, now chugging back toward the land of white high-rises. I tried to read a botanical account of La Gomera, but in the midst of a day so bright white and blue, some crucial, scientific part of my mind seemed not to have made the crossing with me. A steady rattle of wind in the palms hypnotized me into the unthinkable thing for a chronic insomniac: an afternoon nap.

In my sleep I heard a conversation of birds. I woke up and heard it still: birds in the garden, asking each other questions. I shook my cottony head and leaned over the banister, looking down through the trees. From my hidden place I could see only a gardener with bristling white hair. As I spied on him, I saw him thrust a finger into his mouth and make a warbling, musical whistle. In a minute, an answer came back.

I threw on my shoes and hurried downstairs to the garden. It was as deep and edible as Eden: guavas, figs, avocados, a banana tree bent with its burden of fruit. Another tree bore what looked like a watermelon-sized avocado. I located the man I'd seen from up above, but I felt unaccountably shy. I asked him about the tree with outsize avocados, not so much for information as to nurture my fantasy that he would stick his fingers in his mouth and warble the answer. He explained (in Spanish, disappointingly) that the tree comes from Cuba, where they use the fruit as a musical instrument. I asked him to tell me its name in *silbo*. His mouth turned down in a strange pinch and he stood still a long time. Dragonflies clicked in the palms overhead. Finally he said, "She doesn't have a name in *silbo*. She's not from here." And walked off toward the guava trees. A parrot in a wrought-iron cage behind me muttered barely audible Spanish words in monotone; I whistled at him, but he too held me in his beady glare and clammed up.

Breakfast was a sideboard loaded with fresh bread and jars of a sweet something called *miel de palma*, palm honey. I hated to get a house reputation for being nosy, but I was suspicious. It takes both bees and flowers to make honey, and a palm has nothing you'd recognize as a flower. (A botanist with a good eye would, but not a honeybee.) I mentioned this to the cook, who conceded that it's actually not honey but syrup, boiled down from the sap of palms in just the same way New Englanders make syrup from maple sap. I was still suspicious: palm trees, ancient relatives of the grasses, have no xylem and phloem, the little pipes in a tree trunk that move the sap up and down through the tree. To tap a palm, you'd have to whack the head off a mature tree and let it bleed to death.

The cook was congenial about getting a science quiz at this early hour. He told me that what I'd guessed was partly true—in the old days *miel de palma* was a delicacy fatal to the trees, and therefore quite expensive. But in this century, North Africans had developed a gentler palm-tapping technique and introduced it to La Gomera. He said I should go see the palm groves.

I went. I drove up into the highlands of whitewashed villages, vineyards, and deep-cut valleys that rang with the music of wild canaries (ancestral species of their yellower domestic cousins). The island of Gomera is a deeply eroded volcano, twelve miles across and flat-topped, with six mammoth gorges radiating from the center like spokes of a wheel. Farms and villages lie within the gorges, and the road does not go anywhere as the crow flies. Often I rounded a corner to face a stunning view of cliffs and sea and, in the background, the neighboring island of Tenerife. From this distance, Tenerife showed off the pointed silhouette of its own grand volcano, Mount Teide, snowclad from autumn to spring, the highest mountain on any soil claimed by Spain.

La Gomera's farmland brought to mind my grandfather's tales of farming the hills of Kentucky: planting potatoes on ground so steep, he liked to say, you could lop off the ends of the rows and let the potatoes roll into a basket. But here the farmers mostly grow grapevines, on narrow, stone-banked terraces that rise one after another in steep green stairways from coastline to clouds. Hawks wheeled in the air currents rising from the gorges.

I stopped for coffee in a country restaurant that by chance was hosting a family reunion. Unwilling to leave me out, the waiter brought me watercress soup and the country staple known as "wrinkled potatoes." The spicy cilantro sauce had personality. So did the waiter. I told him I'd heard rumors of a village where they make pottery the way the Guanches did. (I was making this up, wholesale.) "Go to Chipude," he said, startling me. "That's not where they make it. The town where they make it doesn't have a name, but you can see it from Chipude."

I followed his advice—how could I not?—and at Chipude they waved me down the road to a place unmarked on my map, but whose residents insist it *does* have a name: Cercado (meaning, approximately, "Hidden inside its walls"). I spotted a group of white-aproned women sitting in an open doorway, surrounded by red clay vessels. One woman wore a beaten straw hat and held a sphere of clay against herself, carving it with a knife. She was not making coils or, technically speaking, building the pot; she was sculpting it, exactly as the Guanches are said to have done. When she tilted up her straw hat, her gold earring glinted and I saw that her eyes were Guanche blue. I asked her where the clay comes from. She pointed with her knife: "That *barranco*," the gorge at the end of the village. Another woman was painting a dried pot with reddish clay slip: mud from that other *barranco*, she pointed. After a pot dries, they explained, and is painted with slip and dries again, its surface is rubbed smooth with a beach rock. Finally, the finished pot is polished to the deep, shiny luster of cherry wood. This last was the task of an old woman with the demeanor of a very old tree, who sat in the corner. She showed me her polishing stick: the worn-down plastic handle of a toothbrush. "What did the Guanches use?" I asked, and she gave me a smile as silent as the gardener's and the parrot's.

The youngest of the women, a teenager named Yaiza, was about to carry a load of finished pots to the kiln. She offered to show me. We walked together through the village, past two girls sitting on the roadside stringing red chilies, down a precarious goat path, into a grassy gorge. The kiln was a mud hut with a tin roof and a serious fire inside. Yaiza adjusted pots on the scorching tin roof, explaining that each one must spend half a day there upside down, half a day right side up, and then it's ready to go into the fire, where it stays another full day. If the weather is right, it comes out without breaking. After this amount of art and labor, the women were prepared to sell one of these pots for about $13. I told Yaiza she should charge ten times that much. She laughed. I asked her if she had ever left La Gomera, and she laughed again, as if the idea were ludicrous. I asked her if a lot of people knew how to make this pottery, and she replied, "Oh, sure. Fourteen or fifteen." All of them belonged to two families, and all lived in this village.

We returned to the pottery house, and I bought a pair of clay bowls. I packed them into my car with enormous care. I had the strange feeling that a few days from now, back at my apartment in Santa Cruz, when I opened these boxes and took out the crumpled newspapers I would possess only air and dust.

The green heart of La Gomera is the Garajonay National Park, a central plateau of ancient laurel forest. On an otherwise dry island, the lush vegetation here drinks from a mantle of perpetual fog. At some point between the dinosaur days and the dawn of humankind, forests like this covered the whole Mediterranean basin; now they have receded to a few green dots on the map in the Madeira and Canary Islands. A friend of mine, an ecologist who studies the laurel forest, had warned me to watch out for sleek black rats in the treetops. At certain seasons the laurels accumulate in their leaves a powerful toxin the rats crave, against their own and the trees' best interests. The local park ranger confirmed this—his advice was to watch for gnawed twigs in the path, then look up, and I would spot the little drug addicts up there. (The colorful Spanish word for drug addiction is *toxico-mania*.) Eventually, he told me, they get so drunk they fall and lie trembling on their backs.

I hiked into the forest, which, like everything else on this island, seemed enchanted. The laurels are old twisted things with moss beards on their trunks and ferns at their feet. Green sunlight fell in pools on the forest floor. I felt drugged myself. I watched the path closely, where I saw tiny white orchids and fallen leaves but no toxicomaniac rats.

Climbing higher, I broke out of the cloud layer into treeless highlands on a bald mountaintop called Pico de Garajonay. The peak is named for the legendary lovers Gara and Jonay, the Guanche equivalents of Romeo and Juliet, who flung themselves with picturesque fatality from this mountaintop. A keen wind whistled over the peak's stone lid, and in the afternoon brightness I could see all but one of the other Canaries. Beyond them to the east lay a long bank of clouds signifying the coast of West Africa. That close. The easternmost Canary Island is only sixty-seven miles from the Saharan sands of mainland Africa. Spain can claim this land all it wants, but geography still asserts itself from time to time, as a reminder that the islands will always belong to Africa. Strong, dry continental storms bring over hot dust and sometimes even torpid, wind-buffeted locusts from the Sahara; Canarians call this dismal weather by its West African name, *kalima*.

But today the air above the clouds was clear as glass, and I felt some electric liquor replacing the blood in my veins. The last time

I looked at that long, pink curving flank of Africa, I was seven years old. I'd sat up all night, thrilled and tightly strung in a Pan Am jet, traveling with my family toward the village in central Congo that would be our home for a time. My father pointed at the cloudbank and told me it was Africa. I couldn't begin to imagine the life that was rolling out ahead of me. But I did understand it would pass over me with the force of a river, and that I needed to pin the water to its banks and hold it still, somehow, to give myself time to know it. I could think of only one way to do it, and I've thought of no better way since. I cracked the spine of the diary I'd received as a Christmas present and began the self-conscious record of my life with this block-lettered sentence:

"When I first saw Africa I thought it was a cloud."

Now I have a desk drawer filled with those diaries, brightly and flimsily bound, with their effete locks and minuscule tin keys. And I have a long bookshelf of the spiral-bound notebooks to which I graduated, once diaries suddenly seemed juvenile. I am still trying to pin the river to its banks.

I left the forest for the dry, windy side of the island, a terrain with the mood of North Africa. Date palms waved like bouquets of feathers. These were the trees tapped for *miel de palma,* and I could see that it doesn't kill the tree outright but it doesn't do it any good, either. The new leaves that spring up after tapping are dwarfed and off kilter. I pulled over, stood beside the road, and looked down into the gorge below, at groves and groves of palms with bad haircuts. The North Africans had saved the trees, but made them ridiculous. The Guanches have survived to whistle a secret life among drug-addicted woodrats. And but for a sneeze of history, Columbus might have stayed forever in the boudoir of Beatriz de Bobadilla. There was nothing at all for me to do about history but write down the wonders that passed over. I felt my mind lift up from its center, unhinge, cast out its months-old plague of despair like locusts into the wind. I laughed out loud.

The shoreline at the base of the gorge was windy, rocky, wild, and utterly deserted. If it wasn't the end of the map, you could surely see it from there. In tide pools, fish and crabs scuttled through their claustrophobic soup, frightened by my long shadow, waiting to be rescued by the next high tide. On the black sand beach I found shells so beautiful I pocketed them with the thrilling sense that I'd stolen something, but there were no witnesses. No one else to see the sun go huge and round, then drown itself, burning a red path of memory on the face of the sea.

In the morning, the air was changed. The garden of the *parador* was quiet, the air choked with pale haze: the *kalima* had come in the night. It deepened its hold as we travelers boarded the ferry and headed back toward an invisible destination. The white-block hotels of southern Tenerife and the giant cone of Mount Teide were nothing, not even mirages in white air. If I had a desk, a home, a life somewhere, it existed only in my mind. When I first saw Africa it was a cloud, and it's surely the same for anything at all. It takes time to peer through the vapor and understand.

As our ferry left the port of San Sebastián, the haze closed down behind us, suspending us on a blank sea between lost worlds. The dolphins were there, I knew. I had written them in my notebook, pinning down the record of my fortune.

CONFESSIONS OF A RELUCTANT ROCK GODDESS

In my hotel room in Boston I sat at the window with my chin on my knees, looking down on the Charles River, where white sails zipped under a freeway overpass, rippling like runaway laundry against a backdrop of late-morning traffic and soot-gray bricks.

I couldn't quite work out what I was doing here, two thousand miles from my daughter, whom I missed so badly I felt as if I'd been shot in the chest, and from my empty Tucson household, where the dust bison roamed freely among the piled-up mail and manuscripts and maybe by now were plotting an unopposed takeover. Instead I was in a hotel, pretending to be a musician on tour with a bunch of authors pretending to be a band.

At the moment I was waiting for two grown men named Hoover and Mouse to come pick up my electric keyboard and haul it to the club in Providence where we would be opening our show that night. Mouse and Hoover were our roadies, hired professionals at my service to tote and tune and do all the dirty work so that I—presumably—could preserve my delicate constitution for the performance. This is a joke; either one of them could play a meaner keyboard than I do, I'm sure, with one or more of his arms in a plaster cast.

I'd asked them to bring the keyboard back here after rehearsal last night, hoping some after-hours practice on my own would render me a passable musician, and then, presto, this very weird scheme would fall into place for me. It didn't. I ran through a halfhearted "Nadine," switched the power off, called my best friend up long distance, and asked: "What in the bejesus am I doing here?" My friend said, as friends do, "I don't know, darlin', but you'll think of something." So far I hadn't. My keyboard was hulking over there on the table like the remains of some malinspired room-service party ordered up at 2 A.M. and left for dead on its tray.

I was suffering from sleep deprivation, that much I knew. I recognized the signs: life seemed baffling and mostly not quite worth the bother. I don't have a musician's sleeping skills, among other things. Our schedule said we were to stay up late rehearsing or performing, then sleep till noon. I've never slept past sunrise I don't think, not on my life, for love nor money nor prescription drugs, so on a schedule like that I had no choice but to stay up till three, get up at six, and sit around my room waiting for scheduled late-morning events like "bag pull" (a new one on me), or the impending visit of Mouse and Hoover.

At last they knocked, and I let them in. They were cheerful and embarrassingly subordinate. "Get those changes worked out?" they asked, as if I'd been pacing all night, frowning in my headphones, memorizing modulations and fingering patterns and stunning new chord sequences.

"You guys are great," I said. It was the truth. They were kind enough to pretend there was work to be done and a show to put on and it's all going to be big fun.

If you ask me, making a fool of yourself on purpose is a scary enterprise. That thought had entered my mind right away, when I first got a letter from Kathi Goldmark asking if I'd be willing to get together with a bunch of other authors and play music for the American Booksellers' Convention in Anaheim. Kathi is a media escort, whose profession involves chauffeuring and nurturing authors when they're on book tours; many of these authors, in the weakened condition induced by too much travel, apparently confessed to her their secret rock-and-roll pasts. Little did we know we would be held to these confessions when Kathi cooked up a scheme. Her form letter offered three boxes to check, suggesting these alternatives: (1) Yes, wild horses couldn't keep me away from a concert in Anaheim! (2) No, I am much too dignified to do such a foolish thing. (3) I might have to wash my hair that night; talk me into it. I checked box #3.

I'm not dignified at all; that wasn't the problem. My friends, under pressure or bribe, will tell stories about me that involve, for example, go-go dancer impersonation and flamboyant petty thievery. (I once helped relocate the Big Boy from his post in front of the Bob's Big Boy establishment to the front porch of an archenemy. The Big Boy weighs a ton.) Dignity has never put any rocks in *my* road. But when I thought over this band idea, it occurred to me that lots of things could keep me away, wild horses being the least of them. I may be fun at parties, but only if I can make it look more or less like an accident; I'm not a *show-off*. To put myself onstage in some kind of crossover talent show seemed audacious.

I received Kathi's recruitment letter several months after she'd mailed it from San Francisco; I was in the Canary Islands, and out of the communicative loop, to put in mildly. After I checked box #3 and dropped my letter into the bright yellow "We-pick-up-mail-when-we-feel-like-it" box down the street from my apartment, I thought that would surely be the end of that.

In February '92, when I moved back to the U.S., a mound of unforwarded mail was waiting for me, shaped something like a faithful dog but much larger. In it I discovered about twenty hot-pink envelopes containing urgent communiqués from Kathi Goldmark. It seemed I was the keyboard player for an all-author band called the Rock Bottom Remainders. Apparently I'd held this position for months. I called Kathi and told her I found it worrisome. Maybe all the other people were first-rate musicians, and I would embarrass them. Or maybe we'd sound hideous. She sent me a tape that Stephen King had made of himself playing guitar and singing. The first of my worries was expunged, and the second, certified.

So we did our crossover talent show, and made a big hit with the tipsy booksellers and publishers of North America, and somebody got the idea we should do it some more. I pointed out that while hit-and-run is one thing, repeat offenders generally get the punishment they richly deserve. If we kept playing, somebody would notice that the Rock Bottom Remainders sounded like Hound Dogs in Heat, with the advantages of modern amplification technology. My fellow band members didn't think this should pose any significant problems.

That winter it became clear that we really were going to do it, something big, possibly a road tour, in the spring of '93. I felt ambivalent. I was finishing a novel, and knew that soon enough I would have to be on the road way too much, promoting the new book. Also, my personal life had become a protracted crisis. I'm a cheerful person on the whole, but '92 was a rotten year. My marriage of many years was transferred suddenly from intensive care to the autopsy table. I had long since come to terms with loneliness, but now I was also going to be single—something I hadn't been since age twenty-two. The death of my family's hopes weighed down my limbs and spirit like a narcotic. Frightening legal demands—which even questioned my ownership of my own writings, in a community property state—left me reeling. And if I wanted to feel sorry for myself on either account, I'd have to work it in between the tasks of being mother of a preschooler, full-time author with impending deadlines, carpool driver, domestic engineer, good citizen, sole breadwinner, and fixer of all broken things around the house. Every single appliance under my roof that involved water—the washing machine, hot-water heater, bathtub, sink, shower, washing machine *again*—chose to blow out and cause a flood on its own special day of that long winter. ("*This*," I announced to my friends, "is a broken home.") Worried sick about cash flow, I tried to fix things myself—with only modest success. The bathtub spigot that I reattached still points skyward, to this day, as though waiting for Mary Poppins to come along and draw a whimsical bath on the ceiling. But Poppins never showed.

I had always watched single working moms with awe, wondering how on earth they did this with no one on standby to help or even cheer them on. Now I was learning. The key is something called "multitasking." You figure out how to combine compatible chores: phone consultations with your editor and washing the breakfast dishes. Writing a novel in the pediatrician's waiting room. Grocery shopping and teaching your child to read. Balancing the budget in the hardware store. Sleeping and worrying. Sobbing and driving.

The notion of a little bus jaunt down the East Coast pretending to be a rock star seemed not so compatible with the other tasks on my list. No way could I do it.

My fellow band members felt otherwise. Ridley Pearson and Dave Barry (bass and lead guitar, respectively), Kathi Goldmark, Tad Bartimus and Amy Tan (vocalists and clotheshorses *par excellence*) all called up to advise me I needed to have some fun. Steve King (rhythm guitar) sent so many mailgrams I became a cult figure at my post office. Roy Blount (band member whose exact contribution remains a mystery) offered to write my novel for me. Throughout that very dreary winter and spring I felt a steady tide of peer pressure and moral support from the Rock Bottom Remainders. In April, when I came home from a long hike on my birthday, my message blinker was having a seizure: every member of the band, I think, had called up to sing "Happy Birthday" into my answering machine. Al Kooper, our musical director and bona fide God of Rock, sang it to the tune of "The Star-Spangled Banner": "*Happy Birth-day to You, from A-al Koo-per . . .*" When he ran low on lyrics, he worked in "*and our flag was still there . . .*" Believe me, I have this tape in a safe place.

Just as a mental exercise, I started working out which friends I would ask to cover the bases at home, if I should ever need to leave for two whole weeks.

In May, I showed up in Boston for the tour.

I'm never nervous at author appearances, I don't care how big the crowd is. I always say, What's the worst that could happen, you're going to forget how to read? Fellow author Richard Nelson once replied in a fierce falsetto, just before we both walked out on stage: "No, you could wet your pants!"

I apologize for my hubris, for I've now known stage fright. The first day I rehearsed with the band in Anaheim I was a case of loose nerve endings in a roomful of people who seemed laced up tight with confidence. In Boston my insecurities were back again with interest. I wanted to play well. Or at least in the right key.

So did everyone else; and it turns out writers are rarely so confident as they seem. Never mind that we created a band persona out of self-ridicule, identifying ourselves publicly as "rhythmically challenged." The truth is, in rehearsal we all paid attention. The famously facetious Dave Barry frowned into space a lot when he played. Ridley sweated and wrote things down. Amy paced. Tad stayed wide-eyed and quiet. Steve made personal breakthroughs. (The second day of rehearsal I told him I thought he was sounding much better. His face lit up like a carnival ride, and he said, "You know what I discovered? When I'm not sure what chord to play, I don't touch the guitar, I just do this—*air strumming!*")

I tried to be dependable and invisible and watch my little synthesizer buttons so I wouldn't come in sounding like a horn section when I was supposed to be an organ, or vice versa. I didn't want Al Kooper to roll his eyes at me. (I've found out since, he rolls his eyes even when he likes you.) I wanted to belong to this gang, and I wasn't going to do it by being the class clown or the silver tongue. We were a whole class of clowns, a league of quick wits, but so what? Can a good pig fly? When we got on stage, we were going to have to be a *band*.

Everybody else developed at least a song or two that was his or her own moment in the spot—Amy had her glorious black-leather "Boots Are Made for Walkin'," Tad embraced "Chain of Fools" with her soul, Ridley committed a righteous "Nadine," Steve excelled (of course) at teenage death songs, and Dave endowed the sixties standard "Gloria" with a new attitude. I supposed I ought to brave center stage too, but the keyboard grows where you plant it, like a tree. It's more of a workhorse than a dance-around-the-stage-and-bite-things type of instrument. Think of one single rock band with a flashy, standout player on boards, if you can. For reasons partly beyond my control, it was very easy for me to fade into oblivion behind Roy Blount's Hawaiian shirt. (In our video, there's no appreciable evidence that I was there.) But bowing to peer pressure, I rashly volunteered to step away from my synthesizer and *sing* "Dock of the Bay." I regretted it instantly.

"Dock of the Bay," Otis Redding version, is my favorite song. To my mind, it speaks to the universal human theme of being washed up somewhere with dashed hopes and poor employment prospects and nobody to hold your hand. I've sung it nine million times in private places, mostly tiled and wet. But I don't sing with my clothes on; it's the principle of the thing. I know my limitations. Or should.

The first time we went through "Dock of the Bay" in rehearsal, my throat was the size of one of those tiny plastic straws they put in your margarita. The guys faithfully played their chords behind my soulfully inaudible rendition, but they examined their sneakers closely when it was over, and I scooted back behind my keyboard like a hermit crab into its shell after a brief interlude of nakedness.

I kind of hoped that song would go away. But Al made me do it again, every day. (He pulled me aside one day and advised that I learn the words. I said, "I know the words, I just can't *sing*.") In time I got the volume up, but not to the point of feeling entitled to sing in front of an audience that had actually paid, in cash, to be there.

This entitlement may or may not have been an issue. The band was philosophically divided on the subject of music. In rehearsal we worked much harder than any of the guys are ever going to admit. We didn't want to embarrass ourselves utterly. But in interviews we knocked each other down in the scramble for the title of Lowest Musical Self-Esteem. It's a face saver. We all knew no amount of rehearsal could ever make us into a first-rate, or even cut-rate, or irate, or reprobate, rock-and-roll band; in that case it's better to pretend you're not trying very hard than to let on that this is really your best effort.

So what was the point, exactly? I found myself brooding a lot, those first early mornings in my Boston hotel room. Why go public with something you know perfectly well you're not doing all that well? Why should good writers play mediocre music? If this is multitasking, I might as well go home and sing "Dock of the Bay" while doing something useful, like banging on the washer.

My rationale, which came to me long after the fact, has to do with a desire to jump fences and graze a lot of pastures, both greener and thornier than the one where I supposedly belong. It looked as if we could raise a huge amount of money to promote literacy, and also I did need a break from an unhappy, hardscrabble time in my life. But those aren't reasons enough. I did it because I want to be exactly what I am—a writer who does other things. Not just a soup-of-the-day double-tasker, Breadwinner Mom;

that's the default option. If I can also be, for one brief moment, Literary Rock Goddess, why not go for broke?

I've spent my life hiding a closetful of other lives. When I entered graduate school in biology in my early twenties, my committee looked long and hard down their noses at my interest in creative writing. And now that I make my way mostly as a writer, it's considered comical or suspect that I have degrees in science. When I speak in public, I'm frequently introduced by someone who will make a point of revealing my checkered past: archaeologist, typesetter, medical technician, translator, biological field researcher, artist's model. The audience generally laughs, and I do too. It seems ridiculous to add music to the list, but it's on there. In 1973, I went to college on a music scholarship. I studied classical piano performance, music theory, and composition at DePauw University for two years, until it occurred to me that all the classical pianists in the U.S. were going to have a shot at, maybe, eleven good jobs, and the rest of us would wind up tinkling through "The Shadow of Your Smile" in a hotel lobby. So I switched to zoology. It seemed practical. I could just as happily have gone over to literature or anthropology or botany. I'm in awe of those people who seem bent from early childhood upon a passionate vocational path. My father, the M.D., tells me that as a first grader he blew up his toy soldiers for the sole purpose of patching them back together. When *I* was a child, if anyone asked what I wanted to be when I grew up, I would reply first of all that I didn't think I would grow up, but on the off chance it happened, I planned to be a farmer and a ballerina and a writer and a doctor and a musician and a zookeeper.

This is not the right answer. I know that now. "Philosopher-king," you might as well say. "Sword swallower–stockbroker. Wrestler–art historian." A business card that lists more than one profession does not go down well in the grown-up set. We're supposed to have one main thing we do well, and it's okay to have hobbies if they are victimless and don't get out of hand, but to confess to disparate passions is generally taken in our society as a sign of attention deficit disorder.

For all the years I studied and worked as a scientist, I wrote poems in the margins of my chemistry texts and field notebooks. But I never identified myself as a poet, not even to myself. It would have seemed self-indulgent. Thoreau was unabashedly both scientific and literary; so was Darwin. But something has happened since then. Life is faster and more streamlined, and there is too much we have to know, just to get the job done right. To get *one* job done right, let alone seven or eight. And certainly we are supposed to get it right.

For all the years I've worked as a writer, I've also played at keyboards and the odd wind instrument, and lately even conga drums. I have sung in the shower. (I sound *great* in the shower.) I have howled backup to Annie Lennox and Randy Travis and Rory Block in my car. I've played in garage bands and jammed informally with musician friends, and with them have even written and recorded a few original songs. But I've *never* called myself a musician. It's not the one thing I do best.

As I get comfortable with the middle stretch of my life, though, it's occurred to me that this is the only one I'm going to get. I'd better open the closet door and invite my other selves to the table, even if it looks undignified or flaky. Possibly this is what's regarded as midlife crisis, but I'm not looking for a new me, just owning up to all the old ones. I *like* playing music. The music I make has not so far been nominated as a significant contribution to our planet, but it's fun.

I've seen those books on multigenre genius: paintings by

Henry James, poetry by Picasso. But I'm not talking about them, I mean the rest of us. I'd like to think it's okay to do a lot of different kinds of things, even if we're not operating at the genius level in every case. I'd like to think we're allowed to have parti-colored days and renaissance lives, without a constant worry over quality control. If the Rock Bottom Remainders were a role model of any kind, I think that was our department: we went on record as half-bad musicians having wholehearted lives.

Thursday night, before our opening show at Shooters Waterfront Café, I bore well in mind the Richard Nelson scenario of What Is the Worst That Could Happen. But that doesn't begin to cover it. You have to picture the whole thing: in our jitters, the men have turned to alcohol and the women to makeup. We have regressed to Girls in the Bathroom mode—sharing hair stuff, asking if this looks okay, relying heavily on each other for fashion advice and kind oversight. This, I imagine, is what other girls did in high school before a big date. I didn't. I skipped the Junior Prom and read Flannery O'Connor. In 1972 I was into blue jeans and defiance, having found that the best defense, where an uninspiring social life was concerned, was a good offense.

My position in this band is ideal: I'm not a Remainder-ette, so I don't do gold lamé and I don't have to be called upon by Al, in rehearsal, as "*Girls!*" At sound check I always tune up with the guys. But on the bus and in the hotel and right now in the dressing room I am definitely *girls.* Lorraine Battle (wardrobe roadie) is giving me a lesson in remedial makeup. I look in the mirror, blink twice as my glamorous big sister smiles back at me. Finally we leave this war-torn dressing room and crowd out onto the backstage bridge, and the guys all hoot at us. I find out what I was missing, in 1972, while I had my nose in a book.

We line up and wait for Roy to introduce us, so that one by one we can run out on the blinding-bright stage and be socked with a roar of cheers. I am invulnerable and supremely transformed: I take the stairs by twos, land onstage in my black lace leggings and long black no-finger gloves, and blow a kiss to the audience. I can't *wait* to sing "Dock of the Bay." I could dance on a table tonight, or roll the Big Boy down the street with impunity. I feel overtly beloved. I lean into my piano and lead out on "Money," and when the bass and guitar kick in I am moving dead center with the In Crowd. I am a river in spring flood season. I may not stop this, ever.

Listen, I could have stayed home and read a book, or plugged earphones into my synthesizer and played "Nadine" to myself, after I put my kid to bed. I almost did. But how many times in your life do you get to be audacious? And really, if you were a kid, would you mind so much if your Girl Scout of a Mom just *once* ran off to be a rock star for two weeks, as long as you got to see the pictures? Think of the ammunition you'd have against her, when your time came.

My daughter thinks it's way cool that I did it. And now that it's over, so do I. The thrill of the Rock Bottom Remainders, for me, was that a crew of mild-mannered writers were audacious *together*. We loved each other for the risks we took, and liked ourselves all right too. I must have sought it out in the middle of my winter, like a seedling straining for sun, because somewhere in my heart's damp basement I knew it's what I needed: Tad's enormous eyes, wide and starry with mascara, smiling at mine in the dressing-room mirror as we prayed we'd hit our notes. Amy in her leather, chin tipped up, glancing over at me for her cue. Steve's little wink when he takes over the whistle reprise on "Dock of the Bay." Dave's grin and Ridley's smiling nod as we look at each other and move, smooth as silk, from A major into the F sharp minor bridge that we *always* screwed up in rehearsal.

Look at us, we are saying to each other. This is really happening. This amazing and joyful noise that has got all those people jammed together and sweating and howling and bumping and grinding is coming from *us*. We are here, right now. We are the ones.

STONE SOUP

In the catalog of family values, where do we rank an occasion like this? A curly-haired boy who wanted to run before he walked, age seven now, a soccer player scoring a winning goal. He turns to the bleachers with his fists in the air and a smile wide as a gap-toothed galaxy. His own cheering section of grown-ups and kids all leap to their feet and hug each other, delirious with love for this boy. He's Andy, my best friend's son. The cheering section includes his mother and her friends, his brother, his father and stepmother, a stepbrother and stepsister, and a grandparent. Lucky is the child with this many relatives on hand to hail a proud accomplishment. I'm there too, witnessing a family fortune. But in spite of myself, defensive words take shape in my head. I am thinking: I dare *anybody* to call this a broken home.

Families change, and remain the same. Why are our names for home so slow to catch up to the truth of where we live?

When I was a child, I had two parents who loved me without cease. One of them attended every excuse for attention I ever contrived, and the other made it to the ones with higher production values, like piano recitals and appendicitis. So I was a lucky child too. I played with a set of paper dolls called "The Family of Dolls," four in number, who came with the factory-assigned names of Dad, Mom, Sis, and Junior. I think you know what they looked like, at least before I loved them to death and their heads fell off.

Now I've replaced the dolls with a life. I knit my days around my daughter's survival and happiness, and am proud to say her head is still on. But we aren't the Family of Dolls. Maybe you're not, either. And if not, even though you are statistically no oddity, it's probably been suggested to you in a hundred ways that yours isn't exactly a real family, but an impostor family, a harbinger of cultural ruin, a slapdash substitute—something like counterfeit money. Here at the tail end of our century, most of us are up to our ears in the noisy business of trying to support and love a thing called family. But there's a current in the air with ferocious moral force that finds its way even into political campaigns, claiming there is only one right way to do it, the Way It Has Always Been.

In the face of a thriving, particolored world, this narrow view is so pickled and absurd I'm astonished that it gets airplay. And I'm astonished that it still stings.

Every parent has endured the arrogance of a child-unfriendly grump sitting in judgment, explaining what those kids of ours really need (for example, "a good licking"). If we're polite, we move our crew to another bench in the park. If we're forthright (as I am in my mind, only, for the rest of the day), we fix them with a sweet imperious stare and say, "Come back and let's talk about it after you've changed a thousand diapers."

But it's harder somehow to shrug off the Family-of-Dolls Family Values crew when they judge (from their safe distance) that divorced people, blended families, gay families, and single parents are failures. That our children are at risk, and the whole arrangement is messy and embarrassing. A marriage that ends is not called "finished," it's called *failed*. The children of this family may have been born to a happy union, but now they are called *the children of divorce*.

I had no idea how thoroughly these assumptions overlaid my culture until I went through divorce myself. I wrote to a friend: "This might be worse than being widowed. Overnight I've suffered the same losses—companionship, financial and practical support, my identity as a wife and partner, the future I'd taken for granted. I am lonely, grieving, and hard-pressed to take care of my household alone. But instead of bringing casseroles, people are acting like I had a fit and broke up the family china."

Once upon a time I held these beliefs about divorce: that everyone who does it could have chosen not to do it. That it's a lazy way out of marital problems. That it selfishly puts personal happiness ahead of family integrity. Now I tremble for my ignorance. It's easy, in fortunate times, to forget about the ambush that could leave your head reeling: serious mental or physical illness, death in the family, abandonment, financial calamity, humiliation, violence, despair.

I started out like any child, intent on being the Family of Dolls. I set upon young womanhood believing in most of the doctrines of my generation: I wore my skirts four inches above the knee. I had that Barbie with her zebra-striped swimsuit and a figure unlike anything found in nature. And I understood the Prince Charming Theory of Marriage, a quest for Mr. Right that ends smack dab where you find him. I did not completely understand that another whole story *begins* there, and no fairy tale prepared me for the combination of bad luck and persistent hope that would interrupt my dream and lead me to other arrangements. Like a cancer diagnosis, a dying marriage is a thing to fight, to deny, and finally, when there's no choice left, to dig in and survive. Casseroles would help. Likewise, I imagine it must be a painful reckoning in adolescence (or later on) to realize one's own true love will never look like the soft-focus fragrance ads because Prince Charming (surprise!) is a princess. Or vice versa. Or has skin the color your parents didn't want you messing with, except in the Crayola box.

It's awfully easy to hold in contempt the straw broken home, and that mythical category of persons who toss away nuclear family for the sheer fun of it. Even the legal terms we use have a suggestion of caprice. I resent the phrase "irreconcilable differences," which suggests a stubborn refusal to accept a spouse's little quirks. This is specious. Every happily married couple I know has loads of irreconcilable differences. Negotiating where to set the thermostat is not the point. A nonfunctioning marriage is a slow asphyxiation. It is waking up despised each morning, listening to the pulse of your own loneliness before the radio begins to blare its raucous gospel that you're nothing if you aren't loved. It is sharing your airless house with the threat of suicide or other kinds of violence, while the ghost that whispers, "Leave here and destroy your children," has passed over every door and nailed it shut. Disassembling a marriage in these circumstances is as much *fun* as amputating your own gangrenous leg. You do it, if you can, to save a life—or two, or more.

I know of no one who really went looking to hoe the harder row, especially the daunting one of single parenthood. Yet it seems to be the most American of customs to blame the burdened for their destiny. We'd like so desperately to believe in freedom and justice for all, we can hardly name that rogue bad luck, even when he's a close enough snake to bite us. In the wake of my divorce, some friends (even a few close ones) chose to vanish, rather than linger within striking distance of misfortune.

But most stuck around, bless their hearts, and if I'm any the wiser for my trials, it's from having learned the worth of steadfast friendship. And also, what not to say. The least helpful question is: "Did you want the divorce, or didn't you?" Did I want to keep that gangrenous leg, or not? How to explain, in a culture that venerates choice: two terrifying options are much worse than none at all. Give me any day the quick hand of cruel fate that will leave me scarred but blameless. As it was, I kept thinking of that wicked third-grade joke in which some boy comes up behind you and grabs your ear, starts in with a prolonged tug, and asks, "Do you want this ear any longer?"

Still, the friend who holds your hand and says the wrong thing is made of dearer stuff than the one who stays away. And generally, through all of it, you live. My favorite fictional character, Kate Vaiden (in the novel by Reynolds Price), advises: "Strength just comes in one brand—you stand up at sunrise and meet what they send you and keep your hair combed."

Once you've weathered the straits, you get to cross the tricky juncture from casualty to survivor. If you're on your feet at the end of a year or two, and have begun putting together a happy new existence, those friends who were kind enough to feel sorry for you when you needed it must now accept you back to the ranks of the living. If you're truly blessed, they will dance at your second wedding. Everybody else, for heaven's sake, should stop throwing stones.

Arguing about whether nontraditional families deserve pity or tolerance is a little like the medieval debate about left-handedness as a mark of the devil. Divorce, remarriage, single parenthood, gay parents, and blended families simply are. They're facts of our time. Some of the reasons listed by sociologists for these family reconstructions are: the idea of marriage as a romantic partnership rather than a pragmatic one; a shift in women's expectations, from servility to self-respect and independence; and longevity (prior to antibiotics no marriage was expected to last many decades—in Colonial days the average couple lived to be married less than twelve years). Add to all this, our growing sense of entitlement to happiness and safety from abuse. Most would agree these are all good things. Yet their result—a culture in which serial monogamy and the consequent reshaping of families are the norm—gets diagnosed as "failing."

For many of us, once we have put ourselves Humpty-Dumpty-wise back together again, the main problem with our reorganized family is that other people think we have a problem. My daughter tells me the only time she's uncomfortable about being the child of divorced parents is when her friends say they feel sorry for her. It's a bizarre sympathy, given that half the kids in her school and nation are in the same boat, pursuing childish happiness with the same energy as their married-parent peers. When anyone asks how *she* feels about it, she spontaneously lists the benefits: our house is in the country and we have a dog, but she can go to her dad's neighborhood for the urban thrills of a pool and sidewalks for roller-skating. What's more, she has three sets of grandparents!

Why is it surprising that a child would revel in a widened family and the right to feel at home in more than one house? Isn't it the opposite that should worry us—a child with no home at all, or too few resources to feel safe? The child at risk is the one whose parents are too immature themselves to guide wisely; too diminished by poverty to nurture; too far from opportunity to offer hope. The number of children in the U.S. living in poverty at this moment is almost unfathomably large: twenty percent. There are families among us that need help all right, and by no means are they new on the landscape. The rate at which teenage girls had babies in 1957 (ninety-six per thousand) was twice what it is now. That remarkable statistic is ignored by the religious right—probably because the teen birth rate was cut in half mainly by legalized abortion. In fact, the policy gatekeepers who coined the phrase "family values" have steadfastly ignored the desperation of too-small families, and since 1979 have steadily reduced the amount of financial support available to a single parent. But, this camp's most outspoken attacks seem aimed at the notion of families getting too complex, with add-ons and extras such as a gay parent's partner, or a remarried mother's new husband and his children.

To judge a family's value by its tidy symmetry is to purchase a book for its cover. There's no moral authority there. The famous family comprised of Dad, Mom, Sis, and Junior living as an isolated economic unit is not built on historical bedrock. In *The Way We Never Were*, Stephanie Coontz writes, "Whenever people propose that we go back to the traditional family, I always suggest that they pick a ballpark date for the family they have in mind." Colonial families were tidily disciplined, but their members (meaning everyone but infants) labored incessantly and died young. Then the Victorian family adopted a new division of labor, in which women's role was domestic and children were allowed time for study and play, but this was an upper-class construct supported by myriad slaves. Coontz writes, "For every nineteenth-century middle-class family that protected its wife and child within the family circle, there was an Irish or German girl scrubbing floors . . . a Welsh boy mining coal to keep the home-baked goodies warm, a black girl doing the family laundry, a black mother and child picking cotton to be made into clothes for the family, and a Jewish or an Italian daughter in a sweatshop making 'ladies' dresses or artificial flowers for the family to purchase."

The abolition of slavery brought slightly more democratic arrangements, in which extended families were harnessed together in cottage industries; at the turn of the century came a steep rise in child labor in mines and sweatshops. Twenty percent of American children lived in orphanages at the time; their parents were not necessarily dead, but couldn't afford to keep them.

During the Depression and up to the end of World War II, many millions of U.S. households were more multigenerational than nuclear. Women my grandmother's age were likely to live with a fluid assortment of elderly relatives, in-laws, siblings, and children. In many cases they spent virtually every waking hour working in the company of other women—a companionable scenario in which it would be easier, I imagine, to tolerate an estranged or difficult spouse. I'm reluctant to idealize a life of so much hard work and so little spousal intimacy, but its advantage may have been resilience. A family so large and varied would not easily be brought down by a single blow: it could absorb a death, long illness, an abandonment here or there, and any number of irreconcilable differences.

The Family of Dolls came along midcentury as a great American experiment. A booming economy required a mobile labor force and demanded that women surrender jobs to return- ing soldiers. Families came to be defined by a single breadwinner. They struck out for single-family homes at an earlier age than ever before, and in unprecedented numbers they raised children in suburban isolation. The nuclear family was launched to sink or swim.

More than a few sank. Social historians corroborate that the suburban family of the postwar economic boom, which we have recently selected as our definition of "traditional," was no panacea. Twenty-five percent of Americans were poor in the mid-1950s, and as yet there were no food stamps. Sixty percent of the elderly lived on less than $1,000 a year, and most had no medical insurance. In the sequestered suburbs, alcoholism and sexual abuse of children were far more widespread than anyone imagined.

Expectations soared, and the economy sagged. It's hard to depend on one other adult for everything, come what may. In the last three decades, that amorphous, adaptable structure we call "family" has been reshaped once more by economic tides. Com- pared with fifties families, mothers are far more likely now to be employed. We are statistically more likely to divorce, and to live in blended families or other extranuclear arrangements. We are also more likely to plan and space our children, and to rate our marriages as "happy." We are less likely to suffer abuse without recourse, or to stare out at our lives through a glaze of prescrip- tion tranquilizers. Our aged parents are less likely to be destitute, and we're half as likely to have a teenage daughter turn up a mother herself. All in all, I would say that if "intact" in modern family-values jargon means living quietly desperate in the bell jar, then hip-hip-hooray for "broken." A neat family model constructed to service the Baby Boom economy seems to be returning gradually to a grand, lumpy shape that human families apparently have tended toward since they first took root in the Olduvai Gorge. We're social animals, deeply fond of companionship, and children love best to run in packs. If there is a *normal* for humans, at all, I expect it looks like two or three Families of Dolls, connected variously by kinship and passion, shuffled like cards and strewn over several shoeboxes.

The sooner we can let go the fairy tale of families functioning perfectly in isolation, the better we might embrace the relief of community. Even the admirable parents who've stayed married through thick and thin are very likely, at present, to incorporate other adults into their families—household help and baby-sitters if they can afford them, or neighbors and grandparents if they can't. For single parents, this support is the rock-bottom definition of family. And most parents who have split apart, however painfully, still manage to maintain family continuity for their children, creating in many cases a boisterous phenomenon that Constance Ahrons in her book *The Good Divorce* calls the "binuclear family." Call it what you will—when ex-spouses beat swords into plowshares and jump up and down at a soccer game together, it makes for happy kids.

Cinderella, look, who needs her? All those evil stepsisters? That story always seemed like too much cotton-picking fuss over clothes. A childhood tale that fascinated me more was the one called "Stone Soup," and the gist of it is this: Once upon a time, a pair of beleaguered soldiers straggled home to a village empty-handed, in a land ruined by war. They were famished, but the villagers had so little they shouted evil words and slammed their doors. So the soldiers dragged out a big kettle, filled it with water, and put it on a fire to boil. They rolled a clean round stone into the pot, while the villagers peered through their curtains in amazement.

"What kind of soup is that?" they hooted.

"Stone soup," the soldiers replied. "Everybody can have some when it's done."

"Well, thanks," one matron grumbled, coming out with a shriveled carrot. "But it'd be better if you threw this in."

And so on, of course, a vegetable at a time, until the whole suspicious village managed to feed itself grandly.

Any family is a big empty pot, save for what gets thrown in. Each stew turns out different. Generosity, a resolve to turn bad luck into good, and respect for variety—these things will nourish a nation of children. Name-calling and suspicion will not. My soup contains a rock or two of hard times, and maybe yours does too. I expect it's a heck of a bouillabaise.

THE SPACES BETWEEN

The drive from Tucson to Phoenix is a trip through merciless desert, where tall saguaros throw up their arms in apparent surrender to the encroaching cotton fields. Some of the land belongs to farmers holding tight to a parched midwestern dream; some belongs to the state of Arizona, mainly because nobody in particular ever bothered to want it. And a big chunk of what we were passing through belongs to the Gila River Reserve, the state's oldest Indian reservation, though nothing I could see from the highway set those particular cacti and irrigated farmlands apart from the rest, as Indian country.

Because Camille was five, and liked to know what to expect at all times, I reminded her that we were on our way to visit the Heard Museum, which was all about Native Americans.

"Indians," I clarified. "You know who Indians are, right?"

"Sure," she said. "People that lived a long time ago."

I felt between my shoulder blades the weight of this familiar frustration. We were driving past fields being tended this very morning, presumably, by Maricopa and Pima Indians. My daughter played routinely with children from other nations including the Tohono O'odham and Yaqui. She had been a guest at their dances and passed almost daily through the Yaqui village that lies between our house and town. But five-year-olds will hear what you tell them, and merrily go right on believing what they *see*. Movies and storybooks say that Indians lived long ago, period, and there's so little else for a modern child to go on.

As a woman with some Cherokee ancestors on my father's side and a blonde, blue-eyed daughter, I find it impossible to pin down the meaning of ethnicity. It's an especially delicate business here in the Southwest, where so many of us boil in one pot without much melting. We're never allowed to forget we are foreign bodies in the eyes of our neighbors. The annual Winter Holiday Concert at Camille's school features a bright patchwork of languages and rituals, each of which must be learned by a different subset of kids, the others having known it since they could talk. It sounds idyllic, but then spend half an hour on the playground and you're also likely to come away with a whole new vocabulary of racial slurs. On the playground no one's counting the strengths of your character, nor the woman your great-grandfather married, unless her genes have dyed your hair and fixed your features. It's the face on your passport that gets you in. Faces that set us apart, in separate houses.

When I pack up my child and head off to a place like the Heard Museum, it's not to claim some piece of our own lost heritage. I have only an inkling of my forebears, and they represent more worlds than I could claim: Scottish stonemasons; Portuguese sailors; farmers from the Eastern Band of Cherokee; planters and sharecroppers and hapless conscripts to both sides of the Civil War. They died without passing on to me the secrets of constructing a limestone chimney flue, navigating by the stars, or planting by the moon. Half the living souls in the southeastern U.S., it seems, claim to be descended from Sacajawea, and that is their business, but I'm not so interested in bloodlines as motivation for multicultural appreciation. I appreciate because I'm interested, just as I can admire tropical fish without being part fish. (And if I *am* part fish, that is *my* business.) We go to the Heard out of love for the great elaborate world, and also to feel more at home in our own neighborhood. I want my child to be so completely familiar with differences that she'll ignore *difference* per se and really see what she's looking at. When she looks at an Acoma water jar, I don't want her to think less of it because it was made by hand in a nonelectrified village high on a mesa. Neither do I want her to think it is the rarefied relic of saints. It seems odd to have to add the latter, but lately we've been besieged with a new, bizarre form of racism that sets apart all things Native American as object of either worship or commerce, depending on your proclivities. It's scary enough to see Kokopelli on a keychain— God for sale, under five dollars—but I'm not much more comfortable with the other angle, the sweat-lodge suburbanites who borrow the material trappings of native ceremonies as a spiritual quickie to offset the stresses of corporate life. What began as anthropology has escalated to fad, and it strikes me that assigning magical power to a culture's every belief and by-product is simply another way of setting those people apart. It's more benign than burning crosses on lawns, for sure, but ultimately not much more humane.

An equal in our time and place is someone with an address and friends, who works and plays and buys groceries in packages with brand names, who is capable of both nobility and mistakes. People who are picture perfect, magical, untouchable, or worse yet, only historic, do not need equal opportunity or educational grants.

An Acoma water jar is just a useful thing, really. Like a soda-pop can, only beautiful.

The Heard Museum stands today because of a hobby that grew out of hand. Dwight and Maie Bartlett Heard settled in the pioneer town of Phoenix in 1895, and long before it was fashionable or provident, they found an absorbing interest in the culture of Arizona's Native peoples. By the 1920s, their collection of artifacts had grown too large and valuable as a community resource to keep on the parlor shelves. Steadily and gently, over more than half a century, the Heard has grown to be one of the world's great centers of Native American heritage.

The entry courtyard welcomed us with the grace of whitewashed arches, orange trees, and weathered *metates*—corn-grinding stones—hunched on the basket-weave brick floor. Mary Brennan, communications coordinator for the museum, met us there, and explained the museum's mission of appreciation for Native people and their culture, especially those of the Southwest. This is not a museum only of artifacts, she pointed out, but of modern Native American life, expressed through both traditional and fine arts. Museum programs bring Native American artists and dancers into schools, for example. Later today there would be a dance performance in the museum auditorium.

I was glad the museum's directors undertook this as part of

their mission: to counter the prevailing notion that Indians made nice pots and shot buffalo and now are dead. I silently wished them luck.

Camille and I were immediately drawn to the wing called "Old Ways, New Ways," a permanent interactive exhibit where kids (and adults, if they're game) can learn to play a drum under the videotaped tutelage of a Kiowa elder, and use a computer to design a Navajo rug, and find enough other adventures to fill an afternoon, easily. I stood with a crew of teenagers at a display showing how the ancient Anasazi fashioned little willow-twig animals that archaeologists frequently find tucked into high crevices in the Grand Canyon. Earnestly we all followed instructions, wrapping and looping our twigs to make horses. Mine looked like a giraffe. I stuffed it deep down in my pocket, wondering if maybe the Anasazi stuck *their* failures into those out-of-the-way crevices for the same reason, and kept the good ones around for the kids to play with.

Camille had better luck fitting wooden forms together to make a Tlingit mural. I stood behind her, watching how two simple shapes—a blunt oval and a curly U-shape—repeat over and over in all the familiar totem-pole aggregations of owl and raven and whale, adding up to that instantly recognizable gestalt of the art of Inuit and other northern tribes. If I hadn't seen it taken apart and reassembled, I would never have understood this amazing principle.

I've always felt half-blind in places where I couldn't touch anything. I find I need to assess textures, and pick things up to see how they're put together; I am far more likely than my child to get in trouble for doing so. Camille has escorted me out of many a china shop. Once, in a Japanese park, I reached out and touched a palace wall because I couldn't identify its material by sight, and wanted to know whether it was stucco or stone; my finger set off great honking alarms and brought a police car up the gravel path. (The lovely signs in Japanese, which I'd taken for part of the decor, apparently said TOUCH THIS AND DIE, HUMBLE TOURIST!) It's true we're a sight-biased species, but still it seems odd that museums that aim to instruct us about a multisensory world tend to convey their information entirely through sight, and maybe a little sound. In such places I generally feel like a child, not quite worthy of the material I'm meant to admire; in the children's wing of the Heard, oddly enough, I felt more respected.

Every part of the museum begged for our attention. The main gallery's permanent collection of ancient and modern Native arts are displayed as a living continuum. The entry is a spare, dark auditorium; in a continuous audiovisual loop, Hopi and Tohono O'odham and Dine people talk directly to the camera about their children and grandparents, their villages, their history, their funerals and blessing ceremonies. Their verbal portraits fall against shifting images of their lives' dramatic backgrounds: the Grand Canyon, Taos Pueblo, saguaros with their arms in the air.

The words of an unidentified Taos Pueblo man are inscribed on the wall of the gallery's entrance: "We have lived upon this land from days beyond history's record, far past any living memory, deep into the time of legend. The story of my people and the story of this place are one single story."

Who else could make this claim? In North America, no one. All American tribes other than the Pueblo have been forced off their home ground, and everyone else migrated here from another hemisphere. The gallery is designed, I think, to stop in our tracks those of us who take transience for granted. It tells an extraordinary tale of human landscapes cradled and shaped by physical ones. Tall photographic murals show the lay of the land,

and the exhibits explain life, history, and survival in these beautiful, severe places. The objects of art in the collection are exquisite, but that is not the point, for all of us have surely seen disembodied pots and baskets in a glass case. Here, those objects lie
together with the matrix of their origins: the colors of Colorado
mud and stone, the need for transporting water, the human passion for both survival and beauty. Baskets that celebrate the whis-
pering colors of grass and the designs of the human heart. Wool
blankets, woven from a pastoral life supported by sheep and a reverence for Spider Woman, the mother of weaving. Blankets so
beautiful they are coveted by people a world away, who can
hardly imagine the sound of bleating sheep in a bone-dry
canyon.

The spaghetti-western caricature of "Indian" had been slip-
ping away from us all day, but it was erased once and for all for
Camille, I think, by the houses. We got to walk into fastidious
replicas of a Zuni pueblo adobe, a Northwest Coast long house,
and a Dine hogan. I've driven many times through the Navajo
reservation in northeastern Arizona and looked longingly at these
low, eight-sided, cozy-looking log hogans, whose chimneys poke
through the center of the roofs to trail thin, blue-gray signals into
the desert sky. I have even stopped by these homes to ask directions, but was never invited in. And now I found one here, dis-
mantled and reassembled in the middle of a gallery. Camille and I
went in and sat on a plank bench with our backs to the hewn
logs, letting our eyes adjust to dimmer light, admiring the way
the home's roundness accommodates both function and the
human need to feel hugged. On the woodstove in the center sat
an iron kettle, waiting (a long time) to cook the next mutton
stew. Camille poked through the assortment of bare necessities
arranged in an open shelf, and touched the traditional velvet

shirts and gathered skirts on coat hangers hung from nails in the wall. She talked as she went, and I was surprised to hear her taking up her own hogan fantasy. "If I meet a Navajo girl in school, maybe she'll invite me home with her and we can sleep on the floor on sheepskins like these."

I got it: my daughter is beginning to believe, truly, that Navajos are people who still walk the earth. They are potential school pals.

Just then, a woman in a sequined sweatshirt ducked in through the doorway, glanced up at the low roof, and remarked before ducking out again, "Boy, they must have been *short* back then."

To write novels, to design a museum, to teach fourth-graders about history—all these enterprises require the interpretation of other lives. And all of them, historically, have been corrupted by privileges of race, class, and gender. The Heard, and places like it, are paddling upstream from the get-go simply by calling themselves "museum." We go there expecting dead things, explained in flat, condescending voices.

"Books," as a category of papery things with the scent of mildew, are paddling up the same stream. I spent plenty of my young womanhood resenting the fact that nearly all the fictional women I'd ever read about were the inventions of men (and that I'd learned about female sexuality from D. H. Lawrence!). But I'm old enough now to stand in the shadow of my former brilliance and face incertitude: would the world really be a better place if Mr. Tolstoy had not invented Anna Karenina, or Mr. Flaubert his Emma Bovary?

More to the point: who, exactly, is entitled to write about the *relationships* between women and men? Hermaphrodites? This is the dilemma upon whose horns I've built my house: I want to know, and to write, about the places where disparate points of view rub together—the spaces between. Not just between man and woman but also North and South; white and not-white; communal and individual; spiritual and carnal. I can think of no genetic or cultural credentials that could entitle a writer to do this—only a keen ear, empathy, caution, willingness to be criticized, and a passionate attraction to the subject.

Of these I can claim in adequate measure only the last; I'm drawn like a kid to mud into the sticky terrain of cultural difference. How wondrous, it seems to me, that someone else can live on the same round egg of a world that I do but explain it differently—how it got here, and what's to be done with it. How remarkable that other people's stories often sound more true to me than my own.

I've been advised from all quarters about my obligations as a writer in the multicultural domain. I have been told explicitly, in fact, both that I should write *more* and *less* (or even *not at all*) about nearly every category of persons imaginable, including men, women, people with disabilities, Asians, Armenians, Native Americans. Fortunately I'm not a short-order cook, because whenever I get lobbed rapid-fire with commands my tendency is to go find a quieter place.

What seems right to me from my quieter place is to represent the world I can see and touch as honestly as I know how, and when writing fiction, to use that variegated world as a matrix for the characters and conflicts I need to fathom. I can't speak in tongues I don't understand, and so there are a thousand tales I'll never tell: the waging of war; coming of age as a man; childhood on an Indian reservation. But when the wounded veteran, the masculine disposition, and the reservation child come into the place where I live, they enter my story. I will watch closely and report on the conversation. A magnificent literary tool is the dramatic point of view; one of its great virtuosos was John Steinbeck. Without ever pretending to know "female" or "Mexican laborer" or "mentally retarded" from the inside, he rendered those characters perfectly from the outside. Through reading Steinbeck I first realized this precious truth: bearing witness is not the same as possession.

Godspeed the right of each of us to speak for ourselves and not be spoken for, but I cannot suffer a possessiveness of stories. When I was nine years old, our town librarian wore broad black picture hats and deeply disliked the idea of children rummaging through her books. I drove her to palsy by checking out every book and dusty pamphlet she had on Cherokee lore, even those she felt God had intended for the Boy Scouts. She told me I would ruin my eyes with so much reading, and hinted my character was headed down the tubes as well. Too late; long before I discovered Cherokee lore, I felt in a certain light that animals could talk. I believed in trees, and that heaven had something to do with how dead trees gentle themselves into long, mossy columns of bright-smelling, crumbling earth, lively inside with sprouting seeds and black beetles. I could not make myself believe in a loud-voiced, bearded God on his throne in the clouds, but I was moved to tears by the compost pile.

No wonder I perturbed the librarian. But her fearful assessment of my soul was inexact. I wasn't studying up to be Cherokee; this would hardly have occurred to me. I loved stories about Wild Boy and the waterbug who discovered the world, not because I wanted to become a different kind of person, but because these stories delighted the heart of the person I already was. And they do still. For my particular brand of pantheism I don't need to affect beads and feathers. I can go to the woods in my jeans and sweatshirt and find grace, without a sweat lodge. I can also fling myself on the floor and spend whole afternoons with my volumes of Joseph Campbell, by accident, when I only meant to be passing by the bookshelf on my way to something productive. I'm not studying up to be Neolithic, I just need those cave paintings and creation stories. I could live without electricity if I had to, but not without stories.

Other people's stories—those are the ones I crave. Not Adam and Eve, designated owners of the garden who get to plunder it and spit it out as they please. Not Noah with his precarious ark, who has set upon us the wrongheaded notion that preserving two specimens of something in a zoo somewhere is all we need of biodiversity. Not the stories I already know, but the ones I haven't heard yet: the ones that will show me a way out of here. The point is not to emulate other lives, or usurp their wardrobes. The point is to find sense. How is a child to find the way to her own beliefs, unless she can stuff her pockets with all the truths she can find—whether she finds them on a library shelf or in a friend's warm, strange-smelling kitchen. The point is for playground slurs to fall dead on her ears, meaningless as locks on an open door. I want to imagine those doors not just open but gone, lying in the dirt, thrown off their hinges by the force of accord in a house of open passage.

Eddie Swimmer stood before us in the auditorium, dressed in moccasins and beaded clothes and a porcupine-hair headdress, explaining the songs and dances. "These songs might all sound to you like 'Hey-ya, hey-ya,' but they're not. Listen. These are words in our languages." Camille and I sat licking our fingers, which were sticky with honey from the Indian fry bread we bought from the concession table at the back. We listened to the singers and watched Eddie do a grass dance, which, in the old days on the plains, had the polite function of stomping down the tall grass before a powwow. Then we watched Derek Davis do the fancy-dance—a fast, difficult type of dancing popular on the modern powwow circuit. Derek's elaborate costume had a beaded breastplate and headdress and showy feather bustles, all put together by members of his family. He pointed out the modern additions: metal bells instead of deer hooves; breechcloths made bright with commercial dyes instead of berries and roots. He was pleased with these improvements, unconcerned about a collector's notion of authenticity. He is a living dancer, a young man in wire-rim glasses and a lot of muscles, definitely not a museum piece. The kids selling fry bread and soft drinks hooted their approval as he began to dance, and when he finished we were all out of breath.

On the way home I asked Camille again, "So, okay, tell me. Who are the Native Americans?"

We'd stayed until closing time, seven hours, a possible world record for museum-going five-year-olds. She spoke sleepily from a horizontal position in the backseat. "They're people who love the earth, and like to sing and dance, and make a lot of pretty stuff to use."

She was quiet for a while, then added, "And I think they like soda pop. Those guys selling the fry bread were drinking a lot of Cokes."

Heaven and earth rejoice. Good enough for now.

POSTCARDS FROM THE IMAGINARY MOM

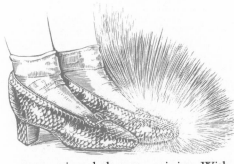

I live for this. Taxiing onto the runway. A craving for adventure afflicts my restless bones like some mineral they are missing. With my sleeve pulled over my palm I rub the airplane window so I'll have a clean view of home falling away underneath me, once we're cleared and my life takes flight.

Oops, better be careful of that sleeve. On this trip it's mandatory that I stay presentable. I'm being sent out on a book tour, four weeks, a different city each day. And for what I'm about to do, I've been given one main piece of advice: Don't check any luggage. If I missed a connection somewhere, my bag would never catch up but would have to follow me from sea to shining sea, one day behind, like a dogged Samsonite version of Lassie Come Home. Better to pare down to the essentials and have nothing to lose. All I have is my mind, and what I'm wearing: sturdy black jeans brand new for this trip, my shiniest cowboy boots, and a nice silk jacket that I hope will pass gracefully from clean, well-lighted bookstores to the Home Shopping Channel. It's a big world out there, so I have a pair of backup shoes in my carry-on: my favorite sneakers, high-tops, red suede.

This is all fine with me—I'm a woman born to travel light. Whatever is coming, I'm ready. We lurch and lift off.

And strangely, for the first time ever, I seize up with airplane phobia. It's not pilot error I dread, but the attendant who's reaching across me to pour coffee. One air pocket, and it could be all over for my silk jacket. I'd have to go home.

Four days out, and I'm hard pressed to remember where I've been. My friends think I'm seeing the U.S.A., but this isn't strictly the case. I'm seeing the inside of bookstores, TV studios, radio stations, newspaper offices, and if I can still see straight at the end of the day, hotel rooms. My spiritual life revolves around the overnight laundry service. I've made these geographic discoveries: all TV studios look exactly alike; all bookstore *bathrooms* look alike; all NPR stations are in the basement.

Not that I'm complaining. A literary novelist whose publisher springs for national promotion has been visited by the angels. For this amazing stroke of luck I vow to feel grateful, and even though my schedule allows no time for exploring, for adventure's sake I'll at least try to spot one major landmark wherever I go. Seattle has Mount Rainier—I looked down on it from the plane. San Francisco, owing to its recent earthquake, has pile drivers going everywhere, a city's dull heartbeat pounding through the subterranean walls of radio stations.

San Diego has fog; on the morning I have to fly from there to L.A. for a live TV show, the airport seems to be closed. I'm getting desperate. If I don't turn up on the set, they may need to interview a potted plant. Suddenly a buzz runs through the airport: *something* is going to L.A. I rush to the counter and miraculously get my ticket changed, my body booked on that plane.

On board, I see this is no miracle, it's only the eight most foolhardy people in San Diego climbing into a prop plane so tiny I'm not allowed to carry my purse on, but must stow it in the hold. I ask the uniformed man, "Will we get breakfast?"

He snorts. "Lady, this flight has a crew of one. You want me to fly the plane, or serve you breakfast?"

We ricochet up through the fog. My fellow travelers blanch, but I relax. Nobody's spilling coffee on *me*.

At the end of my day in L.A., one of my publisher's sales representatives offers to buy me a drink. I accept, though I am so tired I suspect I might be one drink away from delirium. But sales reps are founts of knowledge: they know who's who, how your book is selling, and everything about what's coming out next season. I ask him about an author I've been hearing about— will she be touring?

He avoids my question. "Things happen sometimes," he says. "Not everybody is cut out for the book tour."

"Like what kind of things?"

"Showing up drunk for signings. Punching out a reporter. Going AWOL from the tour, turning up on a shopping spree in Santa Fe. You don't want to know. It's not pretty out there."

I press him, asking again about the famous author in question—does he mean *her*?

"No," he says. "But we decided she's untourable."

Untourable?

Prior to this tour, I went to New York several times to meet editors and publicists over friendly lunches. Were they actually checking to see that my socks matched? These overtures of author-publisher friendship were actually screen tests? I take a deep breath. How ridiculous; I'm thinking like a paranoid schizophrenic.

"What exactly does *untourable* mean?" I ask.

The sales rep stares into his Jack Daniel's and replies, "Insane."

Promoting novels in a sound-bite culture is like selling elephants from a gumball machine. Cramped. Put in your nickel and stand back. Interviewers keep asking, "What is your book *about?*" They mean well. They are kindly giving me a chance to pitch my product. But you should sooner say to a hypochondriac "How are you?" than ask an author this question. Shall I grab you by the lapels and really tell you? Have you got all day? No. What they need is a seven-word answer, and the only accurate one I can think of is: "It's *about* three hundred pages long—*read it*!!" But that sounds surly, so I contrive witty, deficient summaries, which I repeat in senile fashion from city to city.

The words from my own mouth begin to fill me with despair. I'm making a parody of my own earnest trade. If I could say my piece in a glib sentence or two, why on earth would I have spent years of my life on it, and all those pages? If Leo Tolstoy did a book tour for *War and Peace*, how would he answer? "It's about how Napoleon invades Russia, and all these people discover war is, like, bad news." Duh! *Middlemarch* in a plot summary sounds like a soap opera, and *Pilgrim's Progress*, a Sunday-school lesson. My own book doesn't have a prayer in the interview format. I flounder to define not just my own intentions but the concept of novel itself. "It's not so much what *happens*," I try to explain, "but how the words fit together, and what carries over from it into your own life."

My interviewer looks at me, her eyes two perfect asterisks of mascara, and cuts to a commercial.

Through every city, every hour, every question asked and partially answered, I'm missing my daughter. I sleep in an oversized T-shirt she decorated awhile back, with help, in nursery school: it has her picture silk-screened on it, underscored with her name in childish handwriting. But I can't hear her voice on the phone, for I've yet to finish a day and get to a hotel before she's gone to bed. Finally, when I arrive on the East Coast, thanks to the gods of time zones, I can call while she's still awake. At the sound of her small Hello, my heart shudders along my ribcage like a stick dragged down the length of a picket fence.

In a voice much higher-pitched than I remembered it, she details for me her day, the pictures she made, some new kids she met in school. She brightly reports, "I told them I have an imaginary mom."

In Denver, for the noontime news roundup, the commentator clips a mike to my jacket and advises me I'll have fifty-eight seconds to discuss my book. In a flash of insight, I understand everything. In fifty-eight seconds, all I can possibly get across is my name, the color of my jacket, and whether or not I have anything stuck on my teeth. It's not my book that's on sale here. It's me.

Can modern literary success really come down to this, an author's TV persona? In a word, yes. Early on, when a publicist first apprised me of my promotional duties, I whined, "I thought an artist had the privilege of being a recluse!" She firmly replied, "A *starving* artist has that privilege."

An author can say no to a book tour—just as any employee can step backward down the career ladder for the sake of family or peace of mind—but a stigma comes with that choice. From what I've overheard, a writer who won't travel is viewed as an ingrate, a coot, a hermetic unknown who deserves anonymity, or just plain stuck-up. As Garry Trudeau has pointed out, America is the only place where refusal to promote yourself is perceived as arrogance.

Why isn't the author's written word enough? Why must she follow her book out into the world like an anxious mother, to hold its hand and vouch for its character? Why, for that matter, is a book more desirable when it has the author's signature on the flyleaf? I'm so grateful to my readers, heaven knows, I would do anything for them—probably scrub their kitchen floors if they asked. Certainly I would go along peacefully with the book-tour concept, if it were only a matter of my own temporarily disturbed life. But in principle it's an industry trend that worries me. Celebritization of authors rivets the nation's attention on a handful of books each year, shutting out diversity, leaving poets and first novelists to huddle in the cold with the masses of nonfiction scholars whose subject matter is more vital than it is sexy. Readers do need help, of course, in selecting among all the many deserving titles—but what criteria that could possibly fit in a fifty-eight-second TV spot will guide them to an informed choice? The quality of a book's prose means nothing in this race. What will win it a mass audience is the author's ability to travel, dazzle, stake out name recognition, hold up under pressure, look good, and be witty—qualities unrelated, in fact, to good writing, and a lifestyle that is writing's pure nemesis.

What of the brilliant wordsmiths who happen to be elderly, disabled, or indisposed to travel because of young children, or not so great looking, or terribly shy? What are we doing here to the future of literature? Where would we be now if our whole literary tradition were built upon approximately the same precepts as the Miss America competition? Who would win: Eudora Welty or Vanna White?

In Boston I do a syndicated talk show, which I've been told is very important. I'll have eight minutes to explain what my book is about, why everyone should read it, and why I have on these cowboy boots my host keeps staring at. While the makeup person flobs me with a horrid powder puff, I imagine seizing control and turning the tables, interrogating the audience: Why do *you* suppose novelists go on TV? Do you believe in literature? In Tinkerbelle? Clap your hands!

When I'm introduced, my mind rises to the ceiling like an after-death experience and waits up there to see what I'll say this time. I blurt out: "My book is about cowboys and Indians!" This is news to me. I have no idea what it means. For the rest of the interview, one of us, anyway, is on the edge of her seat.

In Atlanta, a talk-show host leans forward just before the cameras roll and confides to me that he's exhausted. "I've had to do two of these shows today, back to back."

"Two shows!" I shout, startling even myself. I left my tact in

San Francisco. "Try *six shows* back to back, plus a couple of readings and book signings and an airplane flight, every single day for three weeks!"

"Oh, but I have the hard part," he tells me sincerely. "I have to sound intelligent."

Apologies to those back home who think I'm lucky, but I've stopped trying to pretend I'm having an adventure. Adventure is stepping through brand-new doors with your mouth shut and your eyes wide open. This is adventure's opposite: traipsing through a hall of mirrors, listening to myself talk. And in truth it's also painfully lonely. I'm surrounded continually by people, good and kind ones, whose appreciation never ceases to astonish me. But I have no control over where, what, and with whom I would like to be. I deeply miss my friends, relaxed conversation, being in a *house*, making myself a sandwich, sitting still with my own thoughts, tucking my child into bed—the things that add up to what matters in my life. I am moving from city to city in a strange glass bubble, the psychic equivalent of that aquarium car that's used for displaying the Pope. Wherever I am, I am there for *now*, and then I will disappear.

In all these days I've smiled at thousands of people, signed their books, and thanked them for their support. Among all those kind strangers, exactly four of them looked me in the eye and said, "You must miss your daughter," or "How long since you've been home?" Each time, tears sprang to my eyes in spite of myself. I am lost somewhere in this crowd. I'm ready to click my heels now and go home.

New York, New York: this has got to be the zenith of my tour. My very publisher himself, along with my agent and the head of publicity, rode with me in the taxi to my reading in Manhattan, and hinted that the three of them would be taking me out afterward for a triumphal celebration. No word on our destination, but sure enough, as we step out of the bookstore just before midnight, we are whisked off in a limo, headed uptown. I feel cheered. How can I complain of a boot-camp schedule when I get treated like royalty in the end?

We pull up to Rockefeller Center. I gawk at the fabulous Deco façade. In the elevator my ears pop on the twentieth floor and again on the fortieth floor as we glide to the top. We are headed for the pinnacle of glamour, the Rainbow Room.

As our excited little party crosses the marble floor, the maître d' approaches us with a polite body block, looks down the full length of his nose, and delivers to us with poetic intonation the sentence of a lifetime:

"I *assume* you are *aware* . . . of our *dress code.*"

We look at each other, bewildered.

"No jeans," he says, "and no sneakers."

I've been wearing this outfit so long I can't imagine the possibility of other clothes, but he's pegged me all right. Jeans. Sneakers. *Suede* sneakers, mind you, but no dice. The maître d' turns to the rest of the party and asks then, "Will there be three of you tonight?"

No one speaks. Lest the congratulations of a thousand fans go to my head, let it be known, I'm a blight on the Rainbow Room.

I consider slinking home in my substandard apparel. Does he realize the alternative was cowboy boots and a T-shirt autographed by a four-year-old? Could his lip get any closer to his nose? Our ambassador of *haute couture* drifts off, leaving us to mortify in the foyer.

In time he returns. And since I have not yet evaporated, he allows regretfully, "It's a slow night tonight. I suppose I could seat you at a back table."

We follow him single file to a back table, from which we are in a position to look down upon the million bright lights of the city. My publisher orders Dom Perignon. "Good," I'm thinking to myself. "We'll show *them* to treat us like pond scum. We'll spend a pile of dough."

But as I toast the town in my jeans and sneakers, my spirits begin to tilt and rise. How is this for poetic justice? I wrote my way to this pinnacle of glamour by one means only: being true to the world I know, a tract of workaday lives where a person is no more likely than, say, a buffalo to rise fifty floors and step out into the hushed terrazzo of the Rainbow Room. My characters could never afford this place—and if by some wild chance they could, they'd probably get scuttled to a back table. That maître d' is no fool. His keen eye caught the *girl* out of which, as they say, *you can't take the country*. And what's wrong with that? If I couldn't be myself, I'd have to be nobody.

Our waiter—bless your soul, wherever you are now—bends down and whispers, "I think you look great."

Thanks. But if I ever go back to the Rainbow Room, I'll be wearing ruby slippers.

I'm nursing a cold, but gloating. I've almost made it. The last stop is a regional booksellers' convention. The plan here is for authors to make an impression on booksellers, who will then go home with a special fondness for us and sell plenty of our books.

All I have to do is give a reading in the morning, then catch a flight home.

I've taken to the behaviors of a stressed laboratory rat—eating furtively in my room, for example, odd things at odd hours. A little past midnight, bleary, sneezing, overdue for bed, I stagger out to the hallway to set out the remains of my room-service tray. The door clicks shut behind me. I don't have my key. I'm standing in the hallway of a finer hotel, wearing an extra-large T-shirt with my daughter's picture on it, and cowboy boots. That's all.

I peck at 1604 and am relieved, when the door opens, to see four ladies in bouffant hairdos having a party in there. They stop talking, arrested by this development at their door: a possible escapee from the Symbionese Liberation Preschool.

As a concise response to everything that has happened in the last month, I begin to sob. I ask one of the ladies if she would dial housekeeping and have them pick up the tray from 1605 and, please, while they're at it, could they bring the key to my room? The ladies do this immediately, since they are not at all keen on the idea of me hanging around crying in their room. They know all they need to know about who I am: namely, that I am deranged.

The next morning, as I give my reading in the convention auditorium, I spot the four ladies in my audience. Turns out they all run bookstores. They are looking at me now as possibly the best story of their lives. If I was sent here to make an impression on booksellers, Lord knows I have done it.

I'm on the plane home, and the devil take my silk jacket. If it's not coffee-stained by the time I get to Tucson, I'll go ahead and have it bronzed. In a few hours I'll hug my little girl. Make dinner. Do laundry. Go to the movies with my friends. Have a life. Return to the work I love, the written word. For the fates and kind readers who allow me to support myself as a writer, may I never forget the height and breadth of my debt. But right at the moment I can't stop thinking of those four booksellers whose party I crashed, and what they will be telling their customers about me. Heaven only knows how the word will spread. Maybe it will get all the way back to the publicity department and the sales reps, by the time I get my next book out. What will they think of me? Maybe that I am . . . this is my wicked thought . . . *untourable*.

THE MEMORY PLACE

This is the kind of April morning no other month can touch: a world tinted in watercolor pastels of redbud, dogtooth violet, and gentle rain. The trees are beginning to shrug off winter; the dark, leggy maple woods are shot through with gleaming constellations of white dogwood blossoms. The road winds through deep forest near Cumberland Falls, Kentucky, carrying us across the Cumberland Plateau toward Horse Lick Creek. Camille is quiet beside me in the front seat, until at last she sighs and says, with a child's poetic logic, "This reminds me of the place I always like to think about."

Me too, I tell her. It's the exact truth. I grew up roaming wooded hollows like these, though they were more hemmed-in, keeping their secrets between the wide-open cattle pastures and tobacco fields of Nicholas County, Kentucky. My brother and sister and I would hoist cane fishing poles over our shoulders, as if we intended to make ourselves useful, and head out to spend a Saturday doing nothing of the kind. We haunted places we called the Crawdad Creek, the Downy Woods (for downy woodpeckers and also for milkweed fluff), and—thrillingly, because we'd once found big bones there—Dead Horse Draw. We caught crawfish with nothing but patience and our hands, boiled them with wild onions over a campfire, and ate them and declared them the best food on earth. We collected banana-scented paw-paw fruits, and were tempted by fleshy, fawn-colored mushrooms but left those alone. We watched birds whose names we didn't know build nests in trees whose names we generally did. We witnessed the unfurling of hickory and oak and maple leaves in the springtime, so tender as to appear nearly edible; we collected them and pressed them with a hot iron under waxed paper when they blushed and dropped in the fall. Then we waited again for spring, even more impatiently than we waited for Christmas, because its gifts were more abundant, needed no batteries, and somehow seemed more exclusively *ours*. I can't imagine that any discovery I ever make, in the rest of my life, will give me the same electric thrill I felt when I first found little righteous Jack in his crimson-curtained pulpit poking up from the base of a rotted log.

These were the adventures of my childhood: tame, I guess, by the standards established by Mowgli the Jungle Boy or even Laura Ingalls Wilder. Nevertheless, it was the experience of nature, with its powerful lessons in static change and predictable surprise. Much of what I know about life, and almost everything I believe about the way I want to live, was formed in those woods. In times of acute worry or insomnia or physical pain, when I close my eyes and bring to mind the place I always like to think about, it looks like the woods in Kentucky.

Horse Lick Creek is a tributary to the Rockcastle River, which drains most of eastern Kentucky and has won enough points for beauty and biological diversity to be named a "wild river." The Nature Conservancy has chosen Horse Lick as a place to cherish particularly, and protect. The creek itself is 16 miles long, with a watershed of 40,000 acres; of this valley, 8,000 acres belong to the Forest Service, about 1,500 to the Nature Conservancy, and the remainder to small farms, whose rich bottoms are given over to tobacco and hay and corn, and whose many steep, untillable slopes are given to forest. The people who reside here have few choices about how they will earn a living. If they are landless, they can work for the school system or county government, they can commute to a distant city, or they can apply for food stamps. If they do have land, they are cursed and blessed with farming. It's rough country. The most lucrative crop that will grow around here is marijuana, and while few would say they approve, everybody knows it's the truth.

Sand Gap, the town at the upper end of the valley, is the straggling remains of an old mining camp. Gapites, as the people of Sand Gap call themselves, take note of us as we pass through. We've met up now with Jim Hays, the Nature Conservancy employee who oversees this holding and develops prospects for purchasing other land to improve the integrity of the preserve. I phoned him in advance and he has been kind enough, on a rainy morning, to show us the way into the preserve. Camille and I jostle in the cab of his pickup like pickled eggs in a jar as we take in the territory, bouncing around blind curves and potholes big enough to swallow at least a good laying hen. We pass a grocery store with a front porch, and the Pony Lot Holiness Church.

JESUS LOVES YOU, BOND WELCOMES YOU, declares a sign in another small settlement.

Jim grew up here, and speaks with the same hill cadences and turns of phrase that shaped my own speech in childhood. Holding tight to the wheel, he declares, "This is the hatefulest road in about three states. Everybody that lives on it wrecks." By way of evidence we pass a rusted car, well off the road and headed down-hollow; its crumpled nose still rests against the tree that ended its life, though it's hard to picture how it got there exactly. Between patches of woods there are pastures, tobacco fields, and houses with mowed yards and flower gardens and folk-loric lawn art. Many a home has a "pouting house" out back, a tarpaper shack where a person can occasionally seek refuge from the rest of the family.

Turner's General Merchandise is the local landmark, meeting place, and commercial hub. It's an honest-to-goodness general store, with a plank floor and a pot-bellied stove, where you can browse the offerings of canned goods, brooms, onion sets, and more specialized items like overalls and cemetery wreaths. A pair of hunters come in to register and tag the wild turkey they've killed—the fourth one brought in today. It's opening day of turkey season, which will last two and a half weeks or until the allotted number of carcasses trail in, whichever comes first. If the season was not strictly controlled, the local turkey population would likely be extinct before first snowfall.

Nobody, and everybody, around here would say that Horse Lick Creek is special. It's a great place to go shoot, drive off-road vehicles, and camp out. In addition to the wild turkeys, the valley holds less conspicuous riches: limestone cliffs and caves that shelter insectivorous bats, including the endangered Indiana bat; shoals in the clear, fast water where many species of rare mussels hold on for their lives. All of this habitat is threatened by abandoned strip mines, herbicide and pesticide use, and literally anything that muddies the water. So earthy and simple a thing as *mud* might not seem hazardous, but in fact it is; fine silt clogs the gills of filter-feeding mussels, asphyxiates them, and this in turn starves out the organisms that depend on the filter feeders. Habitat destruction can be more subtle than a clear-cut or a forest fire; sometimes it's nearly invisible. Nor is it necessarily ugly. Many would argue that the monoculture of an Iowa cornfield is more beautiful than the long-grass prairie that made way for it. But when human encroachment alters the quality of a place that has supported life in its particular way for millions of years, the result is death, sure and multifarious. The mussels of Horse Lick evolved in clear streams, not muddy ones, and so some of the worst offenders here are not giant mining conglomerates but cattle or local travelers who stir up daily mudstorms in hundreds of spots where the road crosses the creek. Saving this little slice of life on earth—like most—will take not just legislation, but change at the level of the pickup truck.

Poverty rarely brings out the most generous human impulses, especially when it comes to environmental matters. Ask a hungry West African about the evils of deforestation, or an unemployed Oregon logger about the endangered spotted owl, and you'll get just about the same answer: I can't afford to think about that right now. Environmentalists must make a case, again and again, for the possibility that we can't afford *not* to think about it. We point to our wildest lands—the Amazon rain forests, the Arctic tundra—to inspire humans with the mighty grace of what we haven't yet wrecked. Those places have a power that speaks for itself, that seems to throw its own grandeur as a curse on the defiler. Fell the giant trees, flood the majestic canyons, and you will have hell and posterity to pay.

But Jackson County, Kentucky, is nobody's idea of wilderness. I wonder, as we bounce along: Who will complain, besides the mute mussels and secretive bats, if we muddy Horse Lick Creek?

Polly and Tom Milt Lakes settled here a hundred years ago, in a deep hollow above the creek. Polly was the county's schoolteacher. Tom Milt liked her looks, so he saved up to buy a geography book, then went to school and asked her to marry him. Both were in their late teens. They raised nine children on the banks of Horse Lick. We pass by their homestead, where feral jonquils mark the ghost-boundaries of a front porch long gone.

Their main visible legacy is the Lakes family cemetery, hidden in a little glade. Camille and I wander quietly, touching headstones where seventy or more seasons of rain have eroded the intentions of permanent remembrance. A lot of babies lie here: Gladys, Colon, and Ollie May Lakes all died the same day they were born. A pair of twins, Tomie and Tiny, lived one and two days, respectively. Life has changed almost unimaginably since the mothers of these children grieved and labored here.

But the place itself seems relatively unaltered—at least at first glance. It wasn't a true wilderness even then, but a landscape possessed by hunters and farmers. Only the contents of the wildcat dumps have changed: the one I stopped earlier to inventory contained a hot-water heater, the headboard of a wooden bed, an avocado-green toilet, a playpen, and a coffee maker.

We make our way on down the valley. The hillside drops steeply away from the road, so that we're looking up at stately maple trunks on the left, and down into their upper branches on the right. The forest is unearthly: filtered light through maple leaves gives a green glow to the creek below us. Mayapples grow in bright assemblies like crowds of rain-slick umbrellas; red trilliums and wild ginger nod from the moss-carpeted banks. Ginseng grows here too—according to Jim, many a young man makes his truck insurance payments by digging "sang."

Deep in the woods at the bottom of a hollow we find Cool Springs, a spot where the rocky ground yawns open to reveal a rushing underground stream. The freshet merely surfaces and then runs away again, noisily, under a deeply undercut limestone cliff. I walk back into the cave as far as I can, to where the water roars down and away, steep and fast. I can feel the cold slabs of stone through the soles of my shoes. Turning back to the light, I see sunlit spray in a bright, wide arc, and the cave's mouth framed by a fringe of backlit maidenhair ferns.

Farther down the road we find the "swirl hole"—a hidden place in a rhododendron slick where the underground stream bubbles up again from the deep. The water is nearly icy and incredibly blue as it gushes up from the bedrock. We sit and watch, surrounded by dark rhododendrons and hemlocks, mesmerized by the repetitious swirling of the water. Camille tosses in tiny hemlock cones; they follow one another in single file along a spiral path, around and around the swirl hole and finally away, downstream, to where this clear water joins the opaque stream of Horse Lick Creek itself.

The pollution here is noticeable. Upstream we passed wildcat strip mines, bulldozed flats, and many fords where the road passes through the creek. The traffic we've seen on this road is recreational vehicles. At one point we encountered two stranded young men whose Ford pickup was sunk up to its doors in what they called a "soup hole," an enormous pothole full of water that looked like more fun than it turned out to be. We helped pull them out, but their engine only choked and coughed muddy water out the tailpipe—not a good sign. When we left them, they were headed back to town on foot.

When Tom Milt and Polly Lakes farmed and hunted this land, their lives were ruled by an economy that included powerful obligations to the future. If the land eroded badly, or the turkeys were all killed in one season, they and their children would not survive. Rarely does any creature have the luxury of fouling its own nest beyond redemption.

But now this territory is nobody's nest, exactly. It's more of a playground. The farmers have mostly gone to the cities for work, and with their hard-earned wages and leisure time they return with off-road vehicles. Careless recreation, and a failure of love for the land, are extracting their pound of flesh from Horse Lick Creek.

A map of this watershed is a jigsaw puzzle of public and private property. The Conservancy's largest holding lies at the lower end of the valley. We pass through Forest Service land to get to it, and park just short of a creek crossing where several tiny tributaries come together. Some of the streams are stained with iron ore, a deep, clear orange. I lean against the truck eating my sandwich while Camille stalks the butterflies that tremble in congregations around the mud puddles—tiger swallowtails. She tries to catch them with her hands, raising a languid cloud of yellow and black. They settle, only mildly perturbed, behind us, as we turn toward the creek.

We make our way across a fallow pasture to the tree-lined bank. The water here is invisibly clear in the shallows, an inviting blue green in the deeper, stiller places. We are half a mile downstream from one of the largest mussel shoals. Camille, a seasoned beachcomber, stalks the shoreline with the delicate thoroughness of a sandpiper, collecting piles of shells. I'm less thrilled than she by her findings, because I know they're the remains of a rare and dying species. The Cumberland Plateau is one of the world's richest sites of mussel evolution, but mussels are the most threatened group in North America. Siltation is killing them here, rendering up a daily body count. Unless the Conservancy acquires some of the key lands where there is heavy creek crossing, these species will soon graduate from "endangered" to "extinct."

Along the creekbanks we spot crayfish holes and hear the deep, throaty clicking of frogs. The high bank across from us is a steep mud cliff carved with round holes and elongated hollows; it looks like a miniature version of the windswept sandstone canyons I've come to know in the West. But everything here is scaled down, small and humane, sized for child adventures like those I pursued with tireless enthusiasm three decades ago. The hay fields beyond these woods, the hawk circling against a mackerel sky, the voices of frogs, the smells of mud and leaf mold, these things place me square in the middle of all my childhood memories.

I recognize, exactly, Camille's wide-eyed thrill when we discover a trail of deer tracks in the soft mud among bird-foot violets. She kneels to examine a cluster of fern fiddleheads the size of her own fist, and is startled by a mourning cloak butterfly (which, until I learned to read field guides, I understood as "morning cloak"). Someone in my childhood gave me the impression that fiddleheads and mourning cloaks were rare and precious. Now I realize they are fairly ordinary members of eastern woodland fauna and flora, but I still feel lucky and even virtuous—a gifted observer—when I see them.

For that matter, they probably *are* rare, in the scope of human experience. A great many people will live out their days without ever seeing such sights, or if they do, never *gasping*. My parents taught me this—to gasp, and feel lucky. They gave me the gift of making mountains out of nature's exquisite molehills. The day I captured and brought home a giant, luminescent green luna moth, they carried on as if it were the Hope diamond I'd discovered hanging on a shred of hickory bark. I owned the moth as my captive for a night, and set it free the next, after receiving an amazing present: strands of tiny green pearls—luna moth eggs—laid in fastidious rows on a hickory leaf. In the heat of my bedroom they hatched almost immediately, and I proudly took my legion of tiny caterpillars to school. I was disappointed when my schoolmates didn't jump for joy.

I suppose no one ever taught them how to strike it rich in the forest. But I know. My heart stops for a second, even now, here, on Horse Lick Creek, as Camille and I wait for the butterfly to light and fold its purple, gold-bordered wings. "That's a morning cloak," I tell her. "It's *very rare*."

In her lifetime it may well be true; she won't see a lot of these butterflies, or fern fiddleheads, or banks of trillium. She's growing up in another place, the upper Sonoran desert. It has its own treasures, and I inflate their importance as my parents once did for me. She signals to me at the breakfast table and we both hold perfectly still, watching the roadrunner outside our window as he raises his cockade of feathers in concentration while stalking a lizard. We gasp over the young, golden coyotes who come down to our pond for a drink. The fragile desert becomes more precious to me as it becomes a family treasure, the place she will always like to think about, after she's grown into adult worries and the need for imaginary refuge.

A new question in the environmentalist's canon, it seems to me, is this one: who will love the *imperfect* lands, the fragments of backyard desert paradise, the creek that runs between farms? In our passion to protect the last remnants of virgin wilderness, shall we surrender everything else in exchange? One might argue that it's a waste of finite resources to preserve and try to repair a place as tame as Horse Lick Creek. I wouldn't. I would say that our love for our natural home has to go beyond finite, into the boundless—like the love of a mother for her children, whose devotion extends to both the gifted and the scarred among her brood.

Domesticated though they are, I want the desert boundary lands of southern Arizona to remain intact. I believe in their remnant wildness. I am holding constant vigil over my daughter's memory place, the land of impossible childhood discovery, in hopes that it may remain a place of real refuge. I hope in thirty years she may come back from wherever she has gone to find the roadrunner thickets living on quietly, exactly as she remembered them. And someone, I hope, will be keeping downy woods and crawdad creeks safe for me.

THE VIBRATIONS OF DJOOGBE

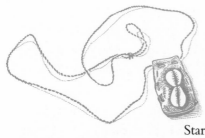

From Benin, West Africa, eight
degrees north of the Equator,
you can see both the North
Star and the Southern Cross. They
crouch above their opposing horizons, ready to guide you north
into the Sahara, or south, down a flank of white beach into the
sea. You can look for them even from Cotonou, Benin's largest
city, where the night sky blazes, untouched as yet by serious
competition from electric lights.

My first night in Benin, my eyes returned again and again to
the sky, searching for my bearings. The night held the tricky, sen-
sual promise of a dream. A wide-bodied jet had touched down
briefly to leave me there, and once it was gone that whole event
of armored comfort seemed as fantastic as extraterrestrial contact. Now I was left to walk through Cotonou's hot, rich-
smelling darkness on streets lined with women selling ordinary

and inconceivable things: grilled bananas, shoes, gasoline sold in liter wine bottles. Each vendor's face was lit by the flame of a small oil-burning lamp; the crowds of tiny lights looked like banks of votive candles in a cathedral, accompanied by a choir of street-smart livestock. Pigeon-sized fruit bats flapped out darkly over the city, beginning their nightly forage.

I walked across a bridge and found the concrete shell of a building that turned out to be a hotel. It looked as if it had been bombed, but that was only a trick of tired eyesight; the building just never got finished. The solemn night clerk showed me to a room with a cot and a sink. He considerately pointed out to me that the door had no lock, and no knob.

At first light, the commerce outside my window rose to fever pitch. The first travelers of the morning were half a dozen small girls driving a herd of pigs; the second, two young men zooming across the bridge on a motorcycle, carrying upright between them a five-foot-square pane of naked glass.

Women crossed the bridge at a more stately pace and moved toward the market balancing gigantic burdens on their heads: bolts of cloth; a mountain of bread; a basket of live chickens with their wings draped over the side, casual as an elbow over the back of a chair. Nearly everyone in Benin dresses in magnificently printed wax cloth, West Africa's trademark garb. Women wrap great rectangles of it around their bodies and heads; men wear it tailored into pajamalike suits or embroidered caftans. The central market is a roar of color, scent, and sound. Next to a pile of dried fish, a tailor works at his open-air table. A woman selling tapioca also does coiffure: a client at her feet can get her hair wrapped with black thread into dozens of pointed, upright sprigs. The market's outskirts grade into industrial zones: cooking fires and small foundries. Beyond this, pigs devoutly work the riverbank garbage dumps.

This is not a country notably equipped for tourism. Every sizable town has at least an inn or two that can provide a bed, a mosquito net, and a decent meal. But mainly Benin is equipped for the business of living as the Beninois do. It's a place to come and witness: to learn, for example, how people use resources when they have no choice but to be resourceful. How armies of little boys at the edge of the market can constitute a city recycling center. You'll have plenty of time to think about everything you've thrown away in your life, as you wind through a labyrinth of palm-frond shelters where hundreds of families are at work, hammering empty oil drums and tomato tins into funnels, buckets, knives, and votive-candle lanterns to light the streets by night.

Not a cubic inch of space in any moving vehicle is wasted. When you flag down a taxi—invariably a subcompact Peugeot— it already has passengers; three in the backseat will make room for a fourth. For thirty cents, you'll get where you need to go.

To travel to a different village you have to take a bush taxi. Go to the particular street corner where trips are organized for your destination. Make an agreement with a driver there, settle on a price. Are you traveling with luggage or animals? These things figure into the cost. Come back in a few hours, the driver will tell you, or tonight—he needs to line up more passengers. You ask, Which? A few hours, or tonight? He says, Both. Maybe you'll make the trip with three other travelers, or five. The driver is in control of this event, and since he is charging by the head, most likely there will be seven. Someone may ride on the roof. Don't worry, it won't be you; you have seniority.

Be prepared to wait. Time is the only thing everyone here has, and they have plenty.

When I told friends I was flying to Benin, alone, with no itinerary, their replies fell into two categories: "Why on earth?" and "Where is that?"

Why on earth is a very good question, though where travel is concerned I'm inclined to let the burden of proof rest in the camp of "Why not?" Among African nations, Benin doesn't have the faunal glamour of Kenya, the cultural cachet of Senegal, nor the political notoriety of Zaire or Somalia. But Africa pulls on me, the whole or any part; having rubbed against it in childhood like iron against a magnet, my poles of attraction are permanently set. Some acquaintances had recently moved to Benin and declared I was welcome to visit. And so, when other work took me as far as London, I stepped off the shelf of Europe into that bewitched place where anything might happen and your French need not be perfect: West Africa.

The second question—where?—is hardly easier to answer. The Republic of Benin, whose name prior to 1975 was Dahomey, passes almost unnoticed on a map: a slim knife of a country between Togo and Nigeria, it is roughly as large and populous as Tennessee. But its narrow borders contain a world of different nations. The climate changes along a north-south gradient from arid savannah down to humid coastal palm plantations, and like most African countries, the modern boundaries reflect colonial decisions that have nothing to do with ethnic unity. Within Benin, and overflowing its borders on all sides, are people who speak Fon, Mina, Yoruba, and other completely unrelated languages. In the northeastern drylands, Islamic influence is strong among the pastoral Fulani. They have little in common with their Somba neighbors, who build castlelike family compounds in the northwest, or the Fon farmers of the south, or the Aiza fishing people who travel by canoe and live in villages of stick houses on stilts over the coastal lagoons.

And so it was that when I asked, in a Cotonou restaurant, for a mashed-yam staple of the north called igname pilé, the waiter grinned broadly and said, "You'll have to come home with me, then. These people down here in the south don't know how to cook."

Southerners are likely to be just as contemptuous of their northern neighbors, who wear startling scars and tattoos on their faces as tribal identifiers. "I would never dream of marrying a woman with tattoos," a Cotonou University student told me, and another young man insisted, when he learned I was going north, that the food in the markets up there is unclean. Members of different tribes, even when they move into the cities, tend to segregate themselves. When the Marxist government led by Ahmed Kerekou—a northerner—was overthrown in 1990, it was on regional grounds as much as ideological ones. Many northerners remain loyal to Kerekou.

In the past, these people had even less in the way of common interest: the Dahomey Kingdom dominated the region for centuries with its army, and amassed stunning wealth by selling the men and women of neighboring tribes into slavery.

Now these tribes, as different as stone, paper, and knife, are crowded into a single national domicile and expected to behave like family; to speak French, agree upon a president, and consider themselves "Beninois." It's a nice theory. The truth is far more interesting.

The 540-kilometer drive to Natitingou is a long, long day. As our bush taxi headed north from Cotonou, commerce gave way to countryside: deep fields of high grass, then forests defoliated from drought, then hillocks of rounded boulders. Termite nests poked up everywhere like gigantic sandcastles. The air hung thick with red dust. It was February, season of the *harmattan*—a hazy heat wave on a languid extended visit from the Sahara. No rain had fallen for four months, and none was expected until late March. Fat-trunked, flat-topped baobab trees punctuated the landscape with comic relief. The car startled a grouse from the roadside brush; the driver swerved, hit it, ran back to collect it. Later we would deliver it to his mother.

Whenever we stopped in a village, which happened often, we were mobbed by children selling bananas. I got out when I could, to walk among the thatch-roofed mud houses, and was greeted by cries of *Yovo! Yovo!*—"white person." I was the first they'd seen all day, maybe all year, and for kids it's a thrilling game. Adults simply say, *"Bonjour, Yovo."* I managed to force a smile, though I felt my pale skin fairly glowing.

"Well, what would you say to an African you saw in America?" a young woman asked, when I complained about this later.

I told her I would not, under any circumstances, say, "Hi, black person."

"Well, here we are all different tribes. We identify ourselves by tribe, and that's how we greet strangers."

I felt faintly consoled, and tried to represent myself—in this land of differences—as a cheerful, upstanding member of the Yovo tribe. Eventually I arrived in Natitingou about the same color as anyone, covered from teeth to shoes in fine red grit. I gratefully showered at the home of my friends, Peace Corps volunteers who taught in Natitingou's secondary school. Their cement house nestled with several others under the canopy of a cashew tree. All night long the apple-sized cashew fruits dropped, socking the tin roof like wayward softballs. Around midnight the bats began to sing in unearthly voices that rang like bells. I lay under my mosquito net, wide awake, unable either to shut away or resist the foreign night.

Northwestern Benin, divided by the dramatic escarpments of the Atakora Chain, is rolling savannah, baobab trees, the Pendjari National Wildlife Refuge, and the remarkable *tatas* of the Somba people. These compounds, scattered out of earshot of one another over the plain, are built of hard red mud like the termite nests and bulge in the same organic way, each one housing an extended family. The cylindrical towers hold stored grain, and high walls connecting them enclose private courtyards. Animals dwell on the ground floor; people sleep upstairs.

I'd been warned that the Somba people are private. But I was fascinated by the lumpy, castlelike *tatas*, and too curious. After visiting the market one day I ambled out across a rutted mealie field, vaguely in the direction of a *tata*. Whistling, I paused to inspect the baobab trees, the ants, the sky, enjoying my nature walk. When I stepped within a stone's throw of the *tata*, an old woman flew out the door, brandishing over her head a yam the size of my arm. I hastened away.

I'd caught only a glimpse of the inner courtyard and its host of fetishes—low mud pedestals crowned with calabash bowls—representing the spirits of ancestors and a conduit to higher powers. Not only the *tatas* but most other villages have fetishes. Usually they appear darkly spattered with fresh blood, a disturbing sight for eyes unaccustomed to such. In Beninois markets I'd seen surly dogs lined up for sale—not as pets. And once along a roadside I caught sight of a procession of young women with live chickens clasped to their heads, dancing toward a ceremonial animal sacrifice.

This part of Africa is the birthplace of *vodoun*, which emigrated with the slave trade to Haiti, Brazil, and other lands where voodoo still thrives. Seventy percent of Beninois place themselves in the category of "animist," where religion is concerned, and nearly everyone wears the *gris-gris,* a personal fetish to ward off bad luck and bad will. It doesn't necessarily preclude belief in Catholicism or Islam; it's simply an acknowledgment of the powers at work here.

My visit happened to coincide with a much-publicized *vodoun* festival, and a pamphlet published for this event explained, in its way, the premise: "Every creature—animal, vegetable, or human, in an obligatory rapport with nature—disposes an energy intermingled with and dependent upon the vibrations of Djoogbe, the most powerful of the vodoun mysteries."

I began to fathom the extent of these mysteries while talking with a man named Julian, who was born in the north but went to Cotonou for a university education. I found him articulate, practical, and by his own assertion, not religious. When we spoke of his family he told me his mother had ten children, of whom five were killed.

I asked, "Five of them died?"

"They were killed," he repeated, pointedly. "My father's other wife was very jealous of my mother."

I was incredulous. "So she *murdered* your brothers and sisters?"

"No, not herself." He was patient with my ignorance. "She went to a fetisher who knew how to use *vodoun*."

Several days later, on the road south again, I kept my eyes on the horizon, where lightning was glancing up like sparks in dry grass. Suddenly the sky broke open and drenched the land. A red flood gushed through village streets, women's draped skirts clung to their legs, and kids danced, ankle-deep. I could not help but point out that there was supposed to be a month left of the dry season. The taxi driver answered flatly, "It's that festival they're having in the south. All those *feticheurs* in one place mess up the weather."

The Royal Palace Museum at Abomey, in central Benin, is a monument made of red clay soil and blood. The twelve successive kings of the Dahomey Empire struck fear through West Africa for two and a half centuries, prior to the French conquest in 1892. When a Dahomey king died, his subjects killed huge numbers of war prisoners in his honor and mixed their blood into the walls of a temple built to house his spirit. The prisoners would otherwise have been sold to Portuguese slave traders, so it's hard to assess the exact degree of their bad luck. I felt chilled, considering their lives, as a museum guide led me through the labyrinth of the palace's red walls. We entered a hall of huge carved animals, the royal icons for different kings: a blue chameleon, a copper-covered lion, a hyena with a poor wide-eyed, half-swallowed goat sticking out of its mouth. (That one, I was told, symbolizes the king's lack of compassion for enemies.) Specifics of history were recorded on giant appliquéd tapestries on the wall. The one devoted to Guézo, ninth in the line of kings, showed Guézo himself engaged in one of his legendary sports: beating an enemy over the head with the unfortunate's own dismembered leg. In the long hall that housed all twelve kings' wooden thrones, Guézo's stood out, twice as high as the others, resting on the skulls of four of his important enemies. "They were Yoruba," the guide stated placidly as I stared at the varnished skulls. "From north of here."

In another courtyard, a small temple held the remains of wives of the tenth ruler, Glélé. When a king died, the guide explained, forty-one of his wives were also killed, to keep him company. ("Would those have been his most or least favorite?" I asked; the guide said, "Probably the prettiest." So. A high price for beauty.) We crossed the compound, past a long row of cannons bought from the Portuguese for fifteen slaves apiece, and arrived at the tomb of Glélé himself. I was asked to remove my sandals, out of respect, before ducking through a doorway into the dim clay room. A fabric-draped bed marked the burial spot. On market days, townspeople bring food here to leave as offerings.

I returned blinking to the bright courtyard, wondering what it could be in this ferocious history that still inspires devotion. African political scientists point out that tribal wars are a legacy of colonialism, with its doctrines of cultural superiority and its habit of roping different cultures together inside arbitrary borders. Undoubtedly this is true, but Abomey stands as a testimony to precolonial horrors. Silently I walked past piles of charred animal bones left behind from a recent ceremony. Yesterday's rain had settled the dust, and in the palms above me even the birds seemed subdued.

Ouidah, the historic point of departure for most of the slaves sold from West Africa, had been chosen as the noisy heart of the International Vodoun Festival. The old Portuguese fort, which houses a museum on slavery, I found crowded with African tourists participating in what mostly resembled a street fair.

On the edge of town, just out of earshot of all the hubbub, was the Sacred Forest, a shadowy glen where fetish chiefs are buried. Huge statues of *vodoun* divinities had been erected there among the trees, for the tourists. I walked among them, stopping to admire Legba, a household protector: he sported the horns of a bull and a huge, erect penis. A woman standing beside me was wearing the image of Pope John Paul II all over her body—a special edition of wax cloth commemorating the recent papal visit. She fanned herself in the steamy heat and rested a hand on Legba's giant bronze foot.

As the only Yovo in sight, I'd attracted a crowd of children. One of them cried whenever I looked at him. Several asked for money, and another wanted to know if I'd marry him. "How many wives are you going to have, total?" I asked. He thought six. "Then forget it," I said.

My entourage and I left the forest and walked through the outskirts of town. I heard a deep boom like a foghorn, then saw, emerging from a doorway, a six-foot, writhing haystack. The children screamed, "*Feticheur!*" and we followed him as he boomed and danced through Ouidah's narrow back streets. In the plaza our fetish joined other dancing haystacks, one of whom had a devil's head. The dancing would go on into the night.

I wandered behind the fort and by pure dumb accident stumbled onto the *vodoun* market. Dozens of fetishers had laid out their wares on the ground: rows of animal skins and bird bodies, turtle skulls, dried chameleons, dark monkey hands lined up in a beseeching row, palms up. I was horrified by this trade in literal flesh and bone (wondering how much of the pharmacopeia was rare or endangered), but also enthralled by the sense of secret business. For nearly an hour I eavesdropped on customers as they recited their maladies and received their prescriptions.

Eventually I collected my nerve and approached the young apprentice of a fetisher. "Something for my love life," I told him. "*Ah oui, mademoiselle*," he said, nodding, with the precise demeanor of a young physician. He introduced himself, asked some diagnostic questions, then led me into a small tent. My heart began to pound. Inside, lined up museumlike, were hundreds of *gris-gris*. There are different types, he explained clinically, for success in business, for improving the memory, for safe travel. He briefly assessed his inventory and produced my love charm: two small sticks bound to a piece of bone, stained dark with blood. This is a powerful one, he said, blessed in a fire ceremony at the temple in Abomey. It has *la force Africaine*. He provided me with extremely complex instructions, promising that if correctly used my *gris-gris* would repel all the wrong sorts of men, attract the right one, and keep him interested. Then he produced a bell from some hidden place, rang it forcefully in my ears, and sang an elaborate chant from which I could pick out only my name, repeated three times. He touched the charm to my collarbone, and then it was mine. For approximately three dollars I walked away with a guarantee of future bliss.

Back at the street fair, in the smoky heat among vendors of souvenirs and street food, a flock of kids danced around a boom box playing Lionel Richie. In the shadow of the Portuguese fort, the history of slavery, and the dark thunderclouds attracted by too many fetishers in one place, this carnival atmosphere struck me as bizarre, if not outright glib. I had a beer at a makeshift café and met a university student, Soulemaine Moreira, whose last name came from the Brazilian family that owned his grandparents. I asked him, "Don't you think about the people who got sent out from this port?"

"We think of them, of course," he replied. "We couldn't forget. Slavery was a terrible tragedy. But look at how it contributed to the cultural development of the New World."

Lionel Richie picked up a new beat, and a woman grilling meat nearby began to dance, elbows out, her fork in the air. Like the Moreira family, this music had made a long, circular trip home. I tried to imagine an America without Michael Jackson or Magic Johnson, without jazz, Motown, break dancing, Rosa Parks, Malcolm X, Taj Mahal, George Washington Carver, James Earl Jones, Maya Angelou. I couldn't picture it—anymore, I'm sure, than Soulemaine Moreira could imagine a Benin without Peugeot taxis. The legacy of colonialism is a world of hurt and cross-pollinated beauty, and we take it from there.

In the gathering dusk I walked through town watching drummers work in tight knots beneath overarching trees, driving their rhythms through crowds who swept with bare feet the dirt floors of these secret amphitheaters. Women moved with babies on their backs. No one kept still.

Finally, very late, I left Ouidah to return to Cotonou in a bush taxi. There were eight of us wedged in. Incredibly, we took on a ninth. In the way of oxygen we had to accept each other's exhalations. Conversations erupted in at least three different languages. I found myself pressed—too tightly to draw a full breath—against the shoulders and thighs of two handsome men. My love charm was burning a hole in my pocket.

It's a hot place, Benin, where everybody has a different story to tell, but every creature has its rapport with nature. It's best to be prepared.

INFERNAL PARADISE

In the darkness before dawn I stood on the precipice of a wilderness. Inches in front of my toes, a lava cliff dropped away into the mammoth bowl of Haleakala, the world's largest dormant volcano. Behind me lay a long green slope where clouds rolled up from the sea, great tumbleweeds of vapor, passing through the pastures and eucalyptus forests of upland Maui to the volcano's crest, then spilling over its edge into the abyss.

Above the rimrock and roiling vapor, the sun was about to break. Far from the world where "Aloha Oe" whines through hotel lobbies, I stood in a remote place at an impossibly silent hour.

But pandemonium had an appointment. Grunting, hissing, a dozen buses pulled up behind me and threw open their doors. Tourists swarmed like ants over the tiny visitors' center at the crater's edge. Loading cameras, dancing from foot to foot in the cold, they positioned for the spectacle. "Darn," a man griped through his viewfinder. "I can't get it all in."

"Take two shots, then," his wife advised.

In the throng I lost and then relocated Steven, my fellow traveler. In his hiking boots, sturdy fedora, and backpack, he apparently struck such a picturesque silhouette against the dawn he'd been cornered by a pro and enlisted as foreground. "Perfect for a wilderness catalog," the photographer testified, while his camera whirred meaningfully.

Sunrise over Haleakala is a packaged Maui tradition: tourists in the beachfront hotels can catch a bus at 3 A.M., ride the winding road to the summit, witness the daybreak moment, and return in time for a late breakfast. As religious experiences go, this one is succinct. In fifteen minutes the crowd was gone.

I wandered a hundred yards back to the parking lot, where a second troop was assembling. For about $120, intrepid sightseers can get a one-way bus ride to the summit for a different thrill: outfitted with helmets, Day-Glo safety vests, and rental bikes, they speed back down to the coast in a huge mob, apparently risking life and limb for a thirty-eight-mile exercise in handbraking. The group leaders, who presumably knew the score, were padded from head to toe like hockey goalies. As they lined up their herds of cyclers, they delivered flat monologues about hand signals and road conditions. "Ready to go play in traffic?" demanded a guide, straddling his mount. "Okay, let's go play in traffic." With the hiss of a hundred thin tires on a ribbon of asphalt, this crowd vanished too.

I blinked in the quiet light, feeling passed over by a raucous visitation. Now the crater lay deserted in the howling wind, by all but one pair of picturesque stragglers. The toes of our boots turned toward the rim and found purchase on a rough cinder trail called Sliding Sands, which would lead us down into the belly of Haleakala. The price: a $6.95 waterproof trail map, and whatever else it might take to haul ourselves down and back again.

Entering the crater at dawn seemed unearthly, though Haleakala is entirely of the earth, and nothing of human artifice. The cliffs absorbed and enclosed us in a mounting horizon of bleak obsidian crags. A lake of cloud slid over the rim, wave by wave, and fell into the crater's separate atmosphere, dispersing in vapor trails. The sharp perimeter of cliffs contains a volcanic bowl three thousand feet deep and eight miles across as the crow flies (or twice that far as the hiker hikes). The depression would hold Manhattan, though fortunately it doesn't.

We walked and slid down miles of gravelly slope toward the crater floor, where the earth had repeatedly disgorged its contents. Black sworls of bubbling lava had once flowed around red cinder cones, then cooled to a tortured standstill. I stood still myself, allowing my eye a minute to take in the lunatic landscape. In the absence of any human construction or familiar vegetation like, say, trees, it was impossible to judge distances. An irregular dot on the trail ahead might be a person or a house-sized boulder. Down below, sections of the trail were sketched across the valley, crossing dark lava flows and green fields, disappearing into a velvet fog that hid the crater's eastern half.

The strange topography of Haleakala Crater makes its own weather. Some areas are parched as the Sahara, while others harbor fern forests under a permanent veil of cloud. Any part of the high-altitude crater can scorch in searing sun, or be lashed by freezing rain, or both, on just about any day of the year. Altogether it is one of the most difficult landscapes ever to host natural life. It is also one of the few places in Hawaii that looks as it did two hundred years ago—or for that matter, two thousand. Haleakala is a tiny, threatened ark.

To learn about the natural history of Hawaii is to understand a story of unceasing invasion. These islands, when they first lifted their heads out of the waves a million years ago, were naked, defiant rock—the most isolated archipelago in the world. Life, when it landed here, arrived only through powerful stamina or spectacular accident: a fern's spore drifting on the trade wind, a seed in the craw of a bird, the bird itself. If it survived, that was an accident all the more spectacular. Natural selection led these survivors to become new species unique in the world: the silversword, for example, a plant that lives in lava beds and dies in a giant flowery starburst; or the nēnē, a crater-dwelling goose that has lost the need for webbed feet because it shuns the sea, foraging instead in foggy meadows, grown languid and tame in the absence of predators. Over the course of a million years, hundreds of creatures like these evolved from the few stray immigrants. Now they are endemic species, living nowhere on earth but here. For many quiet eons they thrived in their sequestered home.

Then humans arrived, also through stamina and spectacular accident. The Polynesians came first, bringing along some thirty plants and animals they considered indispensable, including bananas, taro, sugar cane, pigs, dogs, chickens. And also a few stowaways: rats, snails, and lizards. All of these went forth and multiplied throughout the islands. Each subsequent wave of human immigration brought fresh invasions. Sugar cane and pineapples filled the valleys, crowding out native herbs. Logging operations decimated the endemic rain forests. Pigs, goats, and cattle uprooted and ate whatever was left. Without a native carnivore to stop them, rats flourished like the Pied Piper's dream. Mongooses were imported in a harebrained plan to control them, but the mongoose forages by day and the rat by night, so these creatures rarely encounter one another. Both, though, are happy to feast on the eggs of native birds.

More species have now become extinct in Hawaii than in all of North America. At least two hundred of the islands' endemic plant species are gone from the earth for good, and eight hundred more are endangered. Of the original cornucopia of native birds, many were never classified, including fifty species that were all flightless like the dodo—and now, like the dodo, all gone. A total of only thirty endemic bird species still survive.

It's quite possible now to visit the Hawaiian Islands without ever laying eyes on a single animal or plant that is actually Hawaiian—from the Plumeria lei at the airport (this beloved flower is a Southeast Asian import) to the farewell bouquet of ginger (also Asian). African flame trees, Brazilian jacarandas, mangos and banyans from India, coffee from Africa, macadamia nuts from Australia—these are beautiful impostors all, but to enjoy them is to dance on a graveyard. Exotics are costing native Hawaii its life.

Haleakala Crater is fortified against invasion, because of its protected status as a national park, and because its landscape is hostile ground for pineapples and orchids. The endemics had millennia to adapt to their difficult niche, but the balance of such a fine-tuned ecosystem is precarious, easily thrown into chaos: the plants fall prey to feral pigs and rats, and are rendered infertile by insect invaders like Argentine ants and yellow jacket wasps, which destroy the native pollinators.

Humans have sated their strange appetites in Haleakala too, and while a pig can hardly be blamed for filling its belly, people, it would seem, might know better. The dazzling silverswords, which grow nowhere else on earth, have been collected for souvenirs, leis, scientific study, Oriental medicine, and—of all things—parade floats. These magical plants once covered the ground so thickly a visitor in 1873 wrote that Haleakala's slopes glowed silvery white "like winter in moonlight." But in 1911 a frustrated collector named Dr. Aiken complained that "wild cattle had eaten most of the plants in places of easy access." However, after much hard work he "obtained gunny sacks full." By 1930, it was possible to count the surviving members of this species.

The nēnē suffered an even more dire decline, nearly following the dodo. Since it had evolved in the absence of predators, nothing in this gentle little goose's ground-dwelling habits prepared it for egg-eating rodents, or a creature that walked upright and killed whenever it found an easy mark. By 1951, there were thirty-three nēnē geese left living in the world, half of them in zoos.

Midway through the century, Hawaiians began to protect their islands' biodiversity. Today, a tourist caught with a gunnysack of silverswords would find them pricey souvenirs—there's a $10,000 fine. The Park Service and the Nature Conservancy, which owns adjacent land, are trying to exclude wild pigs from the crater and native forests by means of a fence, though in such rugged ground it's a task akin to dividing needles from haystacks. Under this fierce protection, the silverswords are making a gradual comeback. Nēnē geese have been bred in captivity and reintroduced to the crater as well, but their population, numbered at two hundred and declining, is not yet considered saved. Meanwhile, the invasion creeps forward: even within the protected boundaries of a national park, 47 percent of the plant species growing in Haleakala are aliens. The whole ecosystem is endangered. If the silverswords, nēnē geese, and other colorful endemics of Hawaii survive this century, it will be by the skin of their teeth. It will only happen because we decided to notice, and hold on tight.

Like a child anticipating sleighbells at Christmas, I saw illusory silverswords everywhere. I fixed my binoculars on every shining dot in the distance, and located a lot of roundish rocks in the noonday sun. Finally I saw the real thing. I was not prepared for how they would appear to glow from within, against the dark ground. They are actually silver. For all the world, they look like huge, spherical bouquets of curved silver swords. Cautiously I leaned out and touched one that grew near the path. The knives were soft as bunnies' ears. Unlike the spiny inhabitants of other deserts, the arid-adapted silverswords evolved without the danger of being eaten. Defenseless, they became a delicacy for wild pigs. Such bad luck. This landscape was so unready for what has come to pass.

I never saw "winter in moonlight," but as we trudged deep into the crater we saw silverswords by twos and threes, then clumps of a dozen. Finally, we saw them in bloom. Just once, before dying, the knee-high plant throws up a six-foot flower spike—a monstrous, phallic bouquet of purple asters. If a florist delivered this, you would hide it in a closet. Like a torrid sunset or a rousing thunderstorm, it's the kind of excess that only nature can pull off to rave reviews.

The sun blazed ferociously. My pack was stuffed with a wool sweater, sleeping bag, and rain gear—ludicrous baggage I'd brought at the insistence of Park Service brochures. The gallon of water, on the other hand, was a brilliant idea. The trail leveled out on the valley floor and dusty cinders gave way to fields of delicate-looking ferns, which felt to the touch like plastic. Under a white-hot sky, blue-black cinder cones rose above the fern fields. From the cliffs came the gossipy chatter of petrels, rare endemic gull-like birds that hunt at sea and nest in Haleakala. I envied them their shady holes.

When we topped a small rise, a tin-roofed cabin and water tank greeted us like a mirage. The Park Service maintains a primitive cabin in each of three remote areas of the crater, where hikers, with advance permission, can avail themselves of bunks, a woodstove, and water. (There are no other water sources within Haleakala.) We had a permit for a cabin, but not this one—we would spend the night at Paliku, six miles on down the trail. The next day we would backtrack across the crater by a different path, and exit Haleakala via a formidable set of switchbacks known as the Halemau'u Trail. Even on the level the trail was hard, skulking over knife-edged rocks, requiring exhaustive attention; I could hardly imagine doing this up the side of a cliff. I decided I'd think about that tomorrow.

Meanwhile, we flopped on a grassy knoll at the Kapaloa cabin, devouring our lunchtime rations and most of our water. Steven, my ornithologist companion, observed that we were sitting on a litter of excrement whose source could only be the nēnē. He was very excited about this. I lay down on endangered goose poop and fell asleep.

I woke up groggy, weary of the sun and grateful to be more than halfway to Paliku. We marched through a transition zone of low scrub that softened the lava fields. Ahead of us hung the perpetual mystery of fog that had obscured the crater's eastern end all day, hiding our destination.

Suddenly we walked through that curtain into another world: cool gray air, a grassy meadow where mist dappled our faces and dripped from bright berries that hung in tall briar thickets. We had passed from the mouth of hell to the gates of heaven—presuming heaven looks like the Smoky Mountains or Ireland. Awestruck, and possessed of aching feet, we sat down on the ground. Immediately we heard a quiet honking call. A little zebra-striped goose materialized out of mist and flew very low, circling over our heads. It landed a stone's throw away, cocked its head, and watched us. "Perfect for a wilderness catalog," it might have been thinking. In the past I have scoffed at anthropomorphic descriptions of Hawaii's state bird, which people like to call "friendly" and "curious." Now you can scoff at mine.

Soaked to the bone and suddenly shivering, we walked through miles of deep mist, surrounded by the honking of invisible nēnēs. The world grew quiet, white, punctuated with vermilion berries. The trail ended in Paliku meadow. Beyond the field, a wall of cliffs rose straight up like a Japanese carving of a mountainside in jade. The vertical rock faces were crisscrossed with switchback crevices where gnarled trees and giant ferns sprang out in a sidesaddle forest. On these impossible ledges dwell the last traces of native rain forest. They survive there for only one reason: pigs can't fly.

Paliku cabin, nestled among giant ferns, was a sight for sore muscles. Its iron stove was an antique giant, slow to warm up but ultimately unstoppable. Rain roared on the tin roof of our haven. In the thickening dark we lit candles and boiled water for coffee. I hugged the sleeping bag and heavy wool sweater which, at lunchtime, I'd secretly longed to bury under a rock. It was impossible now to recall the intensity of the morning's heat. And tomorrow I would have trouble believing I'd stood tonight fogging the windowpane with my breath, looking out on the wet tangle of a Hawaiian rain forest. Where does it go when it leaves us, the memory of beautiful, strange things?

At dawn the sun broke over the cliffs and parted the pink mantle of clouds, reaching down like a torch to light the tops of red cinder cones in the crater, one at a time. For half a minute, sunlight twinkled starlike against what must have been the glass front of the visitors' center, all those miles away. I pictured the rowdy scene that must have been playing there. I found I couldn't really believe in any other world but the perfect calm of where I stood.

The mist cleared. Fern trees dripped. Nēnēs flew across the cliff face by twos and threes, in heartbreaking imitation of a Japanese pen-and-ink drawing. Birds called from the trees, leading us on a goose chase through soggy vegetation. We spotted the red *'apapane*, the yellow Maui creeper, and the *'i'iwi*, an odd crimson creature with a downcurved bill—all three gravely threatened species.

I would happily have turned over rocks in search of endangered worms—anything to postpone packing up and striking out. But we had eleven miles to go, all uphill, and the sun was gaining ground. I groaned as I shouldered my pack. "We can still do everything we could when we were twenty," Steven pointed out companionably, "except now it hurts."

We backtracked through the meadows on a trail that grew steadily less muddy. We rested under a crooked acacia, the last tree in an increasingly arid landscape, before taking a new, more northerly trail that would lead us back up and out. Like an old-fashioned hologram, the crater offered two views of itself that were impossible to integrate: all day yesterday we'd walked toward white mist and green cliffs at the crater's wet eastern side; today we did the opposite, facing the drought-stricken western slopes. Planting one boot carefully in front of the other, we crossed acres of black lava flow, where the ground seemed to hula-dance in the heat. We skirted tall cinder cones whose sides were striped yellow and orange like paint pots. Several times I stopped and took note of the fact that there was not, in my whole field of vision, anything living. It might well have been the moon.

The trail graduated from rugged to punishing, and in the afternoon the mists returned. The landscape flowed from lava field to meadow and back again, until we were tossed up at last on the Halemau'u switchbacks. We spent the next two hours scaling the cliff face. With each turn the panorama broadened. We ascended through layers of cloud and emerged on top—nearly two miles above sea level. I invented new names for the Halemau'u trail, which I will keep to myself.

Back home again, still nursing a few aches, I found myself deflecting odd looks from friends who seemed to think a trek through scorched desert and freezing rain in Hawaii was evidence of poor vacation skills.

I would do it all again, in a heartbeat. There are few enough places in the world that belong entirely to themselves. The human passion to carry all things everywhere, so that every place is home, is well on its way to homogenizing our planet. The casualties are the species trampled and lost, extinguished forever, at the rate of tens of thousands per year.

It's a painful, exhausting thing to try to argue logically for the preservation of all the world's species—like trying to debate spirituality with your accountant. Causing extinctions, especially at such a staggering rate, feels dangerous and wrong, but proving scientifically that it's wrong is ultimately very much like proving the existence of God. Commonly environmentalists fall back upon the "pharmacopeia" argument, and it's true enough—any one of these small fallen soldiers might have held some magic bullet to save humanity, like the antirejection drug cyclosporine, derived from a peat-bog fungus, that has made organ transplants a matter of course, or the powerful new anticancer agents extracted from a yew tree. But this seems a pale, selfish reason to care about preserving biodiversity, and near sacrilege in the face of a power so howling and brilliant as life on earth. To love life, really, must mean caring not only for the garden plot but also the wilderness beyond the fence, beauty and mystery for their own sake, because of how meager a world would be without them.

We're familiar enough, across all cultures, with ancestor worship. Why have we never put a second, parallel candle on that altar for "progeny worship"? How can we proceed with such pure disregard for the ones who will come after—not just our own heirs, but all of life? How do we fail to realize we are a point in a grand procession, with equal responsibilities to past and future? "Maybe we need new stories," Linda Hogan writes in the anthology *Heart of the Land*, "new terms and conditions that are relevant to the love of land. . . . We need to reach a hand back through time and a hand forward, stand at the zero point of creation to be certain that we do not create the absence of life, of any species, no matter how inconsequential it might appear to be."

The first tragedy I remember having really understood in my life was the extinction of the dodo. I was four years old. I'd found its picture in the dictionary and asked my mother if we could see a bird like that. I was dismayed by her answer. Not "Yes, at the zoo," or "When you grow up, if you travel to a faraway country."

Just: No. The idea that such a fabulous creature had existed, and then simply stopped being—this is the kind of bad news that children refuse to accept. I hauled the dictionary off to bed with me and prayed for the restoration of the dodo to this earth. I vowed that if I could only see such a creature in my lifetime, I would throw myself in front of its demise.

Haleakala Crater is such a creature in our lifetime. In its great cupped hand it holds a bygone Hawaii, a vision of curled fern leaves, a held-back breath of bird song, things that mostly lie buried now under fields of brighter flowers. The memory of beautiful, strange things slips so far beyond reach, when it goes. If I hadn't seen it, I couldn't care half well enough.

IN THE BELLY OF THE BEAST

The Titans, in the stories of the ancient Greeks, were unearthly giants with heroic strength who ruled the universe from the dawn of time. Their parents were heaven and earth, and their children were the gods. These children squabbled and started a horrific, fiery war to take over ruling the universe.

A more modern legend goes this way: The Titans were giant missiles with atomic warheads. The Pentagon set them in neat circles around chosen American cities, and there they kept us safe and free for twenty-two years.

In the 1980s they were decommissioned. But one of the mummified giants, at least, was enshrined for public inspection. A Titan silo—a hole in the ground where an atomic bomb waited all its life to be launched—is now a missile museum just south of Tucson. When I first heard of it I was dismayed, then curious.

What could a person possibly learn from driving down the interstate on a sunny afternoon and descending into the ground to peruse the technology of nuclear warfare?

Eventually I went. And now I know.

The Titan who sleeps in his sleek, deep burrow is surrounded with ugliness. The museum compound, enclosed by an unkind-looking fence, is set against a lifeless backdrop of mine tailings. The grounds are gravel flatlands. The front office is blank except for a glass display case of souvenirs: plastic hard hats, model missile kits for the kids, a Titan-missile golf shirt. I bought my ticket and was ushered with a few dozen others into a carpeted auditorium. The walls bore mementoes of this silo's years of active duty, including a missile-shaped silver trophy for special achievement at a Strategic Air Command combat competition. The lights dimmed and a gargly voice rose up against high-drama music as the film projector stuttered, then found its stride and began our orientation. A ring of Titan II missiles, we were told, encircled Tucson from 1962 until 1984. The Titan II was "conceived" in 1960 and hammered together in very short order with the help of General Motors, General Electric, Martin Marietta, and other contractors. The launch sites are below ground—"safely protected from a nuclear blast." The missile stands 103 feet tall, 10 feet in diameter, and weighs 150 tons. A fatherly-sounding narrator informed us, "Titan II can be up and out of its silo in less than a minute, hurling its payload at speeds of over 15,000 miles per hour nearly halfway around the world. This ICBM waits quietly underground, its retaliatory potential available on a moment's notice."

The film went on to describe the typical day of a missile crew, and the many tasks required to keep a Titan in a state of constant readiness. Finally we were told sternly, "Little remains to remind people that for 22 years a select group of men stood guard 24 hours a day, seven days a week, protecting the rights and freedom we enjoy in these United States." Day and night the vigilant crew monitored calls from their command post, "Waiting . . . " (a theatrical pause) "for a message that never came."

We filed out of the auditorium and stood in the hostile light of the gravel compound. Dave, our volunteer guide, explained about reinforced antennas that could go on transmitting during an attack (nuclear war disturbs radio transmissions, among other things). One small, cone-shaped antenna sat out in the open where anyone could trip over it. Dave told us a joke: they used to tell the rookies to watch out, this was the warhead. My mind roamed. What sort of person would volunteer to be a bomb-museum docent? The answer: he used to be a commander here. Now, semiretired, he trained cruise-missile operators.

It was still inconceivable that a missile stood erect under our feet, but there was its lid, an enormous concrete door on sliding tracks. Grate-covered holes in the ground bore a stenciled warning: TOXIC VAPORS. During accidents or miscalculations, deadly fuel would escape through these vents. I wondered if the folks living in the retirement community just downhill, with the excruciatingly ironic name of Green Valley, ever knew about this. Dave pointed to a government-issue weathervane, explaining that it would predict which way the poisonous gases would blow. What a relief.

We waited by the silo entry port while a Boy Scout troop emerged. I scanned the little boys' faces for signs of what I might be in for. Astonishment? Boredom? Our group then descended the cool stairwell into the silo. Just like a real missile crew, we put on hard hats to protect ourselves from low-hanging conduits and sharp edges. Signs warned us to watch for rattlesnakes. The hazards of snakes and bumped heads struck me as nearly comic against the steel-reinforced backdrop of potential holocaust. Or, put another way, being protected against these lesser hazards made the larger one seem improbable.

A series of blast doors, each thicker than my body, were all propped open to let us pass. In the old days, you would have had to wait for security clearance at every door in turn before it would admit you and then heave shut, locking behind you. If you turned out to be an unauthorized intruder, Dave explained, you'd get a quick tour of the complex with your face very near the gravel.

Some forty steps down in the silo's bowels, we entered the "No Lone Zone," where at least two people stood guard at all times. This was the control room. Compared with my expectations, undoubtedly influenced by Hollywood, it seemed unsophisticated. The Titan control room was run on cathode-ray tubes and transistor technology. For all the world, it had the look of those fifties spaceship movies, where men in crewcuts and skinny ties dash around trying to figure out what went wrong. No modern computers here, no special effects. The Titan system was built, Dave said, with "we-need-it-now technology." I tried to get my mind around the notion of slapping together some little old thing that could blow up a city.

Dave was already moving on, showing us the chair where the missile commander sat. It looks exactly like a La-z-boy recliner. The commander and one designated enlisted man would have the responsibility of simultaneously turning two keys and engaging the missile, if that call came through. All of us stared mutely at the little holes where those keys would go in.

A changeable wooden sign—similar to the ones the Forest

Service uses to warn that the fire danger today is MEDIUM—hung above the controls to announce the day's STRATEGIC FORCES READINESS CONDITION. You might suppose it went to ultimate-red-alert (or whatever it's called) only a few times in history. Not since the Cuban missile crisis, maybe. You would be wrong. Our guide explained that red-alerts come up all the time, sometimes triggered by a false blip on a radar, and sometimes (unbeknownst to crew members) as a test, checking their mental steadiness. Are they truly sane enough to turn that key and strike up nuclear holocaust? For twenty-two years, every activity and every dollar spent here was aimed toward that exact end, and no other.

"But only the President can issue that order," Dave said. I believe he meant this to be reassuring.

We walked deeper into the artificially lit cave of the silo, down a long green catwalk suspended from above. The entire control chamber hangs on springs like huge shock absorbers. No matter what rocked and raged above, the men here would not be jostled.

On the catwalk we passed an eyewash facility, an outfit resembling a space suit, and a shower in case of mishaps involving toxic missile-fuel vapors. At its terminus the catwalk circled the immense cylindrical hole where the missile stood. We peered through a window into the shaft. Sure enough it was in there, hulking like a huge, dumb killer dog waiting for orders.

This particular missile, of course, is impotent. It has been relieved of its nuclear warhead. Now that the Titans have been decommissioned, they're being used as launch missiles for satellites. A man in our group piped up, "Wasn't it a Titan that blew up a few weeks ago, when they were trying to launch a weather satellite?"

Dave said yes, it was, and he made an interesting face. No one pursued this line of thought, although questions certainly hammered against the roof of my mouth. "What if it'd been headed out of here carrying a payload of death and destruction, Dave, for keeping Tucson safe and free? What then?"

Like compliant children on a field trip, all of us silently examined a metal hatch opening into the missile shaft, through which service mechanics would gain access to the missile itself. A sign on the hatch reminds mechanics not to use their walkie-talkies while inside. I asked what would happen if they did, and Dave said it would totally screw up the missile's guidance system. Again, I felt strangely inhibited from asking very obvious questions: What does this mean, to "totally screw up the missile's guidance system"? That the bomb might then land, for example, on Seattle?

The Pentagon has never discussed it, but the Titan missiles surrounding Tucson were decommissioned, ostensibly, because of technical obsolescence. This announcement came in 1980, almost a decade before the fall of the Berlin Wall; it had nothing to do with letting down the nation's nuclear guard. Make no mistake about this: in 1994 the U.S. sank $11.9 billion into the production and maintenance of nuclear missiles, submarines, and warheads. A separately allocated $2.8 billion was spent on the so-called Star Wars weapons research system. The U.S. government document providing budget authority for fiscal year 1996 states, "Although nuclear forces no longer play as prominent a role in our defense capability as they once did, they remain an important part of our overall defense posture." It's hard to see exactly how these forces are on the wane, as the same document goes on to project outlays of roughly $10 billion for the nuclear war enterprise again the following year, and more than $9 billion every year after that, right on through the end of the century. In Nevada, New Mexico, Utah, Texas, the Great Plains, and many places we aren't allowed to know about, real live atomic bombs stand ready. Our leaders are hard-pressed to pretend some foreign power might invade us, but we are investing furiously in the tools of invasion.

The Pentagon was forced to decommission the Titans because, in plain English, the Titans may have presented one of the most stupendous hazards to the U.S. public we've ever had visited upon us. In the 1960s a group of civilian physicists at the University of Arizona worked out that an explosion at any one of the silos surrounding Tucson would set up a chain reaction among the other Titans that would instantly cremate the city. I learned about this in the late seventies, through one of the scientists who authored the extremely unpopular Titan report. I had months of bad dreams. It was not the first or last time I was floored by our great American capacity for denying objective reality in favor of defense mythology. When I was a child in grade school we had "duck and cover" drills, fully trusting that leaping into a ditch and throwing an Orlon sweater over our heads would save us from nuclear fallout. The Extension Service produced cheerful illustrated pamphlets for our mothers, showing exactly how to stash away in the basement enough canned goods to see the family through the inhospitable aftermath of nuclear war. Now we can pass these pamphlets around at parties, or see the quaint documentary *Atomic Café,* and laugh at the antique charm of such naïveté. And still we go on living in towns surrounded by nuclear choke chains. It is our persistent willingness to believe in ludicrous safety measures that is probably going to kill us.

I tried to exorcise my nightmares in a poem about the Titans, which began:

When God was a child
and the vampire fled from the sign of the cross,
belief was possible.
Survival was this simple.
But the savior clutched in the pocket
encouraged vampires to prosper
in the forest.

The mistake
was to carry the cross,
the rabbit's foot,
the spare tire,
St. Christopher who presides
over the wrecks:
steel cauliflowers
proliferating in junkyard gardens.
And finally
to believe in the fallout shelter.

Now we are left in cities ringed with giants.

Our tour finished, we clattered up the metal stairs and stood once again in the reassuring Arizona sun. Mine tailings on one side of the valley, the pine-crowned Santa Rita mountains on the other side, all still there; beneath us, the specter of hell.

Dave opened the floor for questions. Someone asked about the accident at a Titan silo in Little Rock, Arkansas, where some guy dropped a wrench on the missile and it blew up. Dave wished to point out several things. First, it wasn't a wrench, it was a ratchet. Second, it was a crew of rookies who had been sent in to service the missile. But yes, the unfortunate rookie did drop a tool. It bounced and hit the missile's sheet-metal skin, which is only a quarter of an inch thick. And which doesn't *house* the fuel tank—it *is* the fuel tank. The Titan silo's "blast-proof" concrete lid weighs 740 tons. It was blown 300 yards through the air into a Little Rock cornfield.

Dave wanted us to know something else about this accident: the guys in the shock-absorber-suspended control room had been evacuated prior to the ill-fated servicing. One of them had been drinking a Coke. When they returned they were amazed to see how well the suspension system had worked. The Coke didn't spill.

We crossed the compound to a window where we could look straight down on the missile's nose from above. A woman near me gasped a little. A man asked where this particular missile had been headed for, back in the days when it was loaded, and Dave explained that it varied, and would depend on how much fuel it contained at any given time. Somewhere in the Soviet Union is all he could say for sure. The sight of these two people calmly discussing the specifics of fuel load and destination suddenly scared the living daylights out of me. Discussing that event like something that could really happen. They almost seemed disappointed that it never had.

For years I have wondered how anyone could willingly compete in a hundred-yard dash toward oblivion, and I believe I caught sight of an answer in the Titan museum—in faces that lit up when they discussed targets and suspension systems and megatons. I saw it in eyes and minds so enraptured with technology that they saw before them an engineering spectacle, not a machine designed for the sole purpose of reducing civilizations to rubble.

Throughout the tour I kept looking, foolishly I suppose, for what was missing in this picture: some evidence that the people who ran this outfit were aware of the potential effects of their 150-ton cause. A hint of reluctance, a suggestion of death. In the absence of this, it's easy to get caught up in the internal logic of fuel capacities, circuitry, and chemical reactions. One could even develop an itch to see if this amazing equipment really works, and to measure success in purely technical terms.

The Coke didn't spill.

Outside the silo after the tour, I sat and listened to a young man regaling his girlfriend with further details about the Little Rock disaster. She asked him, "But that guy who dropped the, whatever it was. Did he die?"

The man laughed. "Are you kidding? That door on top was built to withstand a nuclear attack, and it got blown sky-high. Seven hundred and forty tons. That should tell you what happened to the guys inside."

She was quiet for a while, and then asked him, "You really get into that, don't you?"

"Well, sure," he said. "I love machines. It fascinates me what man is capable of designing."

Since that day, I've had the chance to visit another bomb museum of a different kind: the one that stands in Hiroshima. A serene building set in a garden, it is strangely quiet inside, with hushed viewers and hushed exhibits. Neither ideological nor histrionic, the displays stand entirely without editorial comment.

They are simply artifacts, labeled: china saki cups melted together in a stack. A brass Buddha with his hands relaxed into molten pools and a hole where his face used to be. Dozens of melted watches, all stopped at exactly eight-fifteen. A white eyelet petticoat with great, brown-rimmed holes burned in the left side, stained with black rain, worn by a schoolgirl named Oshita-chan. She was half a mile from the hypocenter of the nuclear blast, wearing also a blue short-sleeved blouse, which was incinerated except for its collar, and a blue metal pin with a small white heart, which melted. Oshita-chan lived for approximately twelve hours after the bomb.

On that August morning, more than six thousand school-children were working or playing in the immediate vicinity of the blast. Of most of them not even shreds of clothing remain. Everyone within a kilometer of the hypo-center received more than 1,000 rads and died quickly—though for most of them it was surely not quick enough. Hundreds of thousands of others died slower deaths; many would not know they were dying until two years later, when keloid scars would begin to creep across their bodies.

Every wooden building within two kilometers was annihilated, along with most of the earthquake-proof concrete ones, and within sixteen kilometers every window was smashed. Only concrete chimneys and other cylindrical things were left standing. Fire storms burned all day, creating howling winds and unmeasurable heat. Black rain fell, bringing down radioactive ash, staining walls with long black streaks, poisoning the water, killing fish. I can recite this story but I didn't, somehow, believe it until I looked at things a human being can understand: great handfuls of hair that fell from the head of Hiroko Yamashita, while she sat in her house eight hundred meters from the hypocenter. The pink dress of a girl named Egi-chan, whose blackened pocket held a train ticket out of the city. The charred apron of Mrs. Sato, who was nursing her baby.

The one bizarre, incongruous thing in the museum at Hiroshima, it seemed to me, was a replica of the bomb itself. Dark green, longer than a man, strangely knobbed and finned—it looks like some invention that has nothing to do with people. Nothing at all.

What they left out of the Titan Missile Museum was in plain sight in Hiroshima. Not a sound track with a politically balanced point of view. Just the rest of the facts, those that lie beyond suspension systems and fuel capacity. A missile museum, it seems to me, ought to be horrifying. It had better shake us, if only for a day, out of the illusion of predictability and control that cradles the whole of our quotidian lives. Most of us—nearly all, I would say—live by this illusion. We walk through our days with our minds on schedule—work, kids, getting the roof patched before the rainy season. We do not live as though literally everything we have, including a history and a future, could be erased by two keys turning simultaneously in a lock.

How could we? How even to pay our monthly bills, if we held in mind the fact that we are camped on top of a technological powder keg? Or to use Carl Sagan's more eloquent analogy: we are all locked together in a room filled with gasoline vapors, insisting that because *they* have two hundred matches, *we* won't be safe until we have *three* hundred.

The Cold War is widely supposed to have ended. But preparations for nuclear war have not ended. The Titan museum's orientation film is still telling the story we have heard so many times that it sounds, like all ultra-familiar stories, true. The story is that *they* would gladly drop bombs on us, if they weren't so scared by the sheer toughness of our big missiles. *They* are the aggressors. *We* are practicing "a commitment to deterrence."

Imagine you have never heard that story before. Look it in the eye and see what it is. How do strategic-games trophies and Titan-missile golf shirts stack up against a charred eyelet petticoat and handfuls of hair? The United States is the only nation that has ever used an atomic bomb. Dropped it, on men and women and schoolchildren and gardens and pets and museums, two whole cities of quotidian life. We did it, the story goes, to hasten the end of the war and bring our soldiers home. Not such an obvious choice for Oshita-chan. "To protect the rights and freedoms we enjoy" is a grotesque euphemism. Every nuclear weapon ever constructed was built for the purpose of ending life, in a manner so horrific it is nearly impossible to contemplate. And U.S. nuclear science has moved steadily and firmly, from the moment of its birth, toward first-strike capacity.

If the Titan in Green Valley had ever been allowed to do the job for which it was designed, the fire storm wouldn't have ended a world away. Surely all of us, even missile docent Dave, understand that. Why, then, were we all so polite about avoiding the obvious questions? How is it that a waving flag can create an electromagnetic no-back-talk zone? In 1994, half a century after the bombing of Hiroshima, we spent $150 billion on the business and technology of war—nearly a tenth of it specifically on nuclear-weapons systems. Any talk of closing down a military base raises defensive and reverent ire, no matter how wasteful an installment it might be. And yet, public debate dickers and rages over our obligation to fund the welfare system—a contribution of about $25 a year from each taxpayer on average, for keeping the poorest among us alive. How can we haggle over the size of this meager life preserver, while shiploads of money for death sail by unchallenged? What religion of humankind could bless the travesty that is the U.S. federal budget?

Why did I not scream at the top of my lungs down in that hole?

I didn't, so I'll have to do it now, to anyone with the power to legislate or listen: one match in a gasoline-filled room is too many. I don't care a fig who is holding it.

I donned the hard hat and entered the belly of the beast, and I came away with the feeling of something poisonous on my skin. The specter of that beast could paralyze a person with despair. But only if you accept it as inevitable. And it's only inevitable if you are too paralyzed with despair to talk back. If a missile museum can do no more than stop up our mouths, with either patriotic silence or desperation, it's a monument the living can't afford. I say slam its doors for good. Tip a cement truck to the silo's gullet and seal in the evil pharaoh. If humanity survives long enough to understand what he really was, they can dig him up and put on display the grandiose depravity of the twentieth century.

I left, drove down into the innocent palm-shaded condominiums of Green Valley, and then, unexpectedly, headed up the other side of the valley into the mountains. When I reached the plateau of junipers and oaks I pulled off the road, hiked into the woods, and sat for a long time on a boulder in the middle of a creek. Water flowed away from me on either side. A canopy of sycamore leaves whispered above my head, while they waited for night, the close of one more day in which the world did not end.

In a poem called "Trinity," Sy Margaret Baldwin explained

why she would never go down to the site of the first atomic-
bomb explosion, which is opened to the public every year:

> . . . I would come face to face with my sorrow, I
> would feel hope slipping from me and be afraid
> the changed earth would turn over and speak
> the truth to the thin black ribbons of my ribs.

JABBERWOCKY

Once upon a time, a passing stranger sent me into exile. I was downtown in front of the Federal Building with a small crowd assembled to protest war in the Persian Gulf; he was in a black Ford pickup. As the truck roared by he leaned most of his upper body out the window to give me a better view of his finger, and he screamed, "Hey, bitch, love it or leave it!"

So I left.

He wasn't the first to give me that instruction; I've heard it since I was a nineteen-year-old in a scary barbershop haircut. Now I was thirty-four, mother of a child, with a decent reputation and pretty good hair. Why start listening *now*? I can only say he was finally one too many. I was on the verge of having a special kind of nervous breakdown, in which a person stalks through a Kmart parking lot ripping yellow ribbons off car antennas.

I realize that would have been abridging other people's right to free expression. What was driving me crazy was that very term "right to free expression," and how it was being applied in a nation at war. We were supposed to behave as though we had refrigerators for brains. Open, shove in a slab of baloney, close, stay cool. No questions. Our leaders told us this was a *surgical* war. *Very clean.* The language of the event was a perfect construct of nonmeaning. "Delivering the ordnance," they called it on the nightly news, which sounds nearly friendly . . . "Why, here is your ordnance, friends, just sign on the line." "Deliver the ordnance" means "Drop the bomb."

But we bought the goods, or we kept our mouths shut. If we felt disturbed by the idea of pulverizing civilizations as the best way to settle our differences—or had trouble explaining that to our kids as adult behavior—we weren't talking about it. Typically, if I raised the debate, I was advised that if I liked Saddam so much I could go live in Iraq. As a matter of fact I *didn't* like Saddam, *or* the government of Kuwait. The two countries appeared practically indistinguishable; I doubt if many Americans could have guessed, a few years earlier (as we flooded Iraq with military aid), which one would turn out to be the Evil Empire, and which would require us to rush to its defense in the name of democracy. If *democracy* were really an issue we considered when going into that war, Iraq might have come out a nose ahead, Kuwait being a monarchy in which women held rights approximately equal to those of livestock. (*Since* the war, women's status in Kuwait has actually declined.) But the level of discourse allowed on this subject was "We're gonna kick butt." A shadow of doubt was viewed as treason.

I'm lucky enough to have a job that will follow me anywhere, so I left. I could contemplate from a distance these words on patriotism, written by the wise Garry Wills: "Love of one's country should be like love of one's spouse—a give-and-take criticism and affection. Although it is hoped one prefers one's spouse to other people . . . one does not prove that one loves one's wife by battering other women."

Give-and-take criticism and affection, out the window. And the battery was severe. Upon moving to Spain I read in the papers what was common knowledge, apparently, everywhere but in the U.S.: from the first night onward we bombed Iraqis relentlessly in their homes, killing thousands of civilians every day. Within months, more than 250,000 would be dead—most of them children—because of bombed-out water and sewer systems, hospitals with no antibiotics, hospitals with no roofs. To my horror I read that infections of hands and feet were rampant among Iraqi children, because of bombing debris, and the only available treatment was amputation. It had been an air war on civilians. The Commission of Inquiry for the International War Crimes Tribunal is still compiling the gruesome list of what the United States bombed in Iraq: all the country's major dams and most of its drinking water facilities; enough sewage treatment facilities to contaminate the Tigris River with waterborne killers; virtually all communications systems, leaving civilians unwarned of danger and unable to get help; civilian cars, buses, and taxis; 139 auto and railway bridges; food-processing, storage, and distribution systems; 100 percent of irrigation systems; wheat and grain fields (with incendiary bombs); 28 civilian hospitals and 52 community health centers; clothing factories; a cosmetics factory; an infant formula factory; 56 mosques; more than 600 schools. This was our surgical war.

Soon after the bombing ended, Ramsey Clark wrote a book

called *The Fire This Time*, a meticulously researched account of the many ways the U.S. violated the Geneva Convention and perpetrated crimes against civilians in the Persian Gulf War. Clark, as a former U.S. Attorney General, had once been appointed trustee of the nation's conscience. Now he asked us to reckon with some awful responsibilities. But he encountered a truly American form of censorship: free enterprise in the hands of a monkey called See No Evil. His manuscript was rejected by eleven publishers—every major New York house. The editors did not turn it down for lack of merit, they said, but on grounds that it wouldn't be popular. (At length it was released by a small publisher called Thunder's Mouth; hurray for the alternative presses.)

No such hard luck for the memoirs of generals or celebrities, or O. J. Simpson's thoughts from jail while awaiting his verdict. The publisher of the latter (Little, Brown) claimed no moral qualms about providing a forum for Simpson at a time when he already commanded more media attention than has ever been held, probably, by any human being on the planet. The first printing was half a million copies.

This is a spooky proposition: an information industry that narrows down what we'll get to read and know about, mainly on the basis of how eagerly we'll lap it up. Producers and publishers who make these choices seem inclined, if confronted, to throw up their hands and exclaim, "I can't help it if that's what the people want!" A mother could say the same while feeding her baby nothing but jelly beans day after day; so could a physician who administers morphine for head colds. Both would be convicted of criminal neglect. Why is there no Hippocratic Oath for the professionals who service our intellects? Why is it that I knew, without wanting to, every possible thing about a figure skater who got whacked on the leg with a pipe—a melodrama that in the long run, let's face it, is utterly without consequence to anyone but the whackers and the whackee—but I had to go far out of my way to dig up the recent historical events that led to anarchy in Somalia and Haiti? (I learned, it's worth noting, that the U.S. did embarrassing things in both places.) News stations will move heaven and earth to get their own reporters into the likes of California vs. O. J. Simpson, or backstage with Tonya Harding, but not into hearings on the Clean Air Act. Producers will blame consumers, but blame is hardly the point if we are merrily dying of ignorance, and killing others with our apathy. Few U.S. citizens are aware, for example, that our government has routinely engineered assassinations of democratically elected heads of state in places like Chile and Guatemala, and replaced them with such monstrous confederates as Augusto Pinochet and Castillo Armas. Why do those dictators' names fail even to ring a bell in most red-blooded American heads? Possibly because our heads are too crowded with names like O. J. and Tonya. The guilt for that may not rest entirely with the producers or the consumers, but the crime has nevertheless occurred. To buy or to sell information as nothing more than a consumer product, like soda pop, is surely wrong. Marketed in that way, information's principal attribute must be universal palatability.

This is not to say we only get to tune in to *happy* news—there are wrecks and murders galore. But it's information that corroborates a certain narrow view of the world and our place in it. Exhaustive reports of rare, bizarre behaviors among the wealthy support the myth that violent crime is a random, unpreventable disaster, and obscure the blunt truth that most crime is caused by poverty. There's not much in the news to remind us, either, that poverty is a problem we could decently address, as all other industrialized countries have done. The safest marketing technique is to dispense with historical analysis, accountability, and even—apparently—critical thought.

When the Smithsonian deferred to what it called "public pressure" and canceled an exhibit on the historical use of the atomic bomb in Hiroshima and Nagasaki, Smithsonian Secretary I. Michael Heyman explained, "Veterans and their families were expecting, and rightly so, that the nation would honor and commemorate their valor and sacrifice. They were not looking for analysis, and, frankly, we did not give enough thought to the intense feeling that such analysis would evoke." *Analysis* in that case meant the most elementary connection between cause and effect: what happens when the Ordnance gets Delivered.

As a member of that all-important public, I'd like to state for the record that I'm offended. Give me the chance and I'll spend my consumer dollar on the story that relates to what kind of shape the world will be in fifty years from now. I'll choose analysis, every time, over placebo news and empty salve for my patriotic ego. I'm offended by the presumption that my honor as a citizen will crumple unless I'm protected from knowledge of my country's mistakes. I'm made of sturdier stuff than that, and I imagine, if he really thought about it, so is that guy who leaned out of a truck to give me the finger. What kind of love is patriotism, if it evaporates in the face of uncomfortable truths? What kind of honor sits quietly by while a nation's conscience flies south for a long, long winter?

Artists are as guilty as anyone in the conspiracy of self-censorship, if they succumb to the lure of producing only what's sure to sell. The good ones don't, and might still sell anyway, for humans have long accepted subconsciously that good art won't always, so to speak, match the sofa. "Poets are the unacknowl-edged legislators of the race," Percy Shelley said. They are also its margin of safety, like the canaries that used to be carried into mines because of their sensitivity to toxic gases; their silence can be taken as a sign of imminent danger.

The artist's maverick responsibility is sometimes to sugarcoat the bitter pill and slip it down our gullet, telling us what we did-n't think we wanted to know. But in the U.S. we're establishing a modern tradition of tarpapering our messengers. The one who delivers the bitter pill, whether the vehicle is a war-crime docu-mentary or a love story, is apt to be dismissed as a "political artist."

It's a Jabberwockish sort of label, both dreaded and perplex-ing. Technically the term "political" refers to campaigns, govern-ments, and public institutions. But *Police Academy* was not called political. Barry Lopez is called political, and he writes about dying ecosystems and great blue herons and wolves, for heaven's sake. It took me years to work out what it is that earns this scald-ing label for an artist or an act.

Now I know, and this is how I know: during the Gulf War some young friends of mine wanted to set up a table in the shop-ping mall and hand out information about the less cheerful aspects of the war. The administrators of the mall refused permission. My friends contended, "But you let people hand out yellow ribbons and flags and 'We kick butt' bumper stickers!" The mall adminis-trators explained their charter forbids anything political. "Handing out yellow ribbons is public service," they said, "but what *you* want to do is *political.*"

Now you know. This subterfuge use of the word "political," which doesn't show up in my Random House Unabridged, means only that a thing runs counter to prevailing assumptions. If

60 percent of us support the war, then the expressions of the other 40 percent are political—and can be disallowed in some contexts for that reason alone. The really bad news is that the charter of the shopping mall seems to be standing in as a national artistic standard. Cultural workers in the U.S. are prone to be bound and gagged by a dread of being called political, for that word implies the art is not quite pure. Real art, the story goes, does not endorse a point of view. This is utter nonsense, of course (try to imagine a story or a painting with no point of view), and also the most thorough and invisible form of censorship I've ever encountered. When I'm interviewed about writing, I spend a good deal of time defending the possibility that such things as environmental ruin, child abuse, or the hypocrisy of U.S. immigration policy are appropriate subjects for a novel. I keep waiting for the interviewer to bring up *art* things, like voice and metaphor; usually I'm still waiting for that when the cows come home.

In rural Greece some people believe that if you drink very cold water on a very hot day, you will die; here, we have that kind of superstition about mixing art with conscience. It's a quaintly provincial belief that fades out fast at our borders. Most of the rest of the world considers social criticism to be, absolutely, the most legitimate domain of art. If you think I'm overstating this, look who's been winning Nobel Prizes in literature for the last ninety years:

Nadine Gordimer, who has spent her life writing against racism and apartheid in South Africa. Joseph Brodsky, who spent some years in Siberia because of his criticism of Soviet society. Wole Soyinka, who has also logged time in jail because of his criticisms of colonialism in Africa. Gabriel García Márquez, who is possibly the most gifted social critic in a whole continent of social-critic-writers. Czeslaw Milosz, who was active in the anti-Nazi underground and whose poetry is thoroughly ideological. Pablo Neruda, Aleksandr Solzhenitsyn, Miguel Asturias, Thomas Mann, George Bernard Shaw.

U.S. prizewinners do not dominate this list (as they do the Nobel categories of Physics, Chemistry, and Medicine), especially since the 1950s. It's not for lack of great writers, but perhaps because we've learned to limit our own access to serious content. The fear of being perceived as ideologues runs so deep in writers of my generation it undoubtedly steers us away from certain subjects without our knowing it. The fear is that if you fall short of perfect execution, you'll be called "preachy." But falling short of perfection when you've plunged in to say what needs to be said—is that so much worse, really, than falling short when you've plunged in to say what *didn't* need to be said?

And if you should by chance succeed—oh, then. Art has the power not only to soothe a savage breast, but to change a savage mind. A novel can make us weep over the same events that might hardly give us pause if we read them in a newspaper. Even though the tragedy in the newspaper happened to real people, while the one in the novel happened in an author's imagination.

A novel works its magic by putting a reader inside another person's life. The pace is as slow as life. It's as detailed as life. It requires you, the reader, to fill in an outline of words with vivid pictures drawn subconsciously from your own life, so that the story feels more personal than the sets designed by someone else and handed over via TV or movies. Literature duplicates the experience of living in a way that nothing else can, drawing you so fully into another life that you temporarily forget you have one of your own. That is why you read it, and might even sit up in bed till early dawn, throwing your whole tomorrow out of whack, simply to find out what happens to some people who, you know perfectly well, are made up. It's why you might find yourself crying, even if you aren't the crying kind.

The power of fiction is to create empathy. It lifts you away from your chair and stuffs you gently down inside someone else's point of view. It differs drastically from a newspaper, which imparts information while allowing you to remain rooted in your own perspective. A newspaper could tell you that one hundred people, say, in an airplane, or in Israel, or in Iraq, have died today. And you can think to yourself, "How very sad," then turn the page and see how the Wildcats fared. But a novel could take just one of those hundred lives and show you exactly how it felt to be that person rising from bed in the morning, watching the desert light on the tile of her doorway and on the curve of her daughter's cheek. You would taste that person's breakfast, and love her family, and sort through her worries as your own, and know that a death in that household will be the end of the only life that someone will ever have. As important as yours. As important as mine.

At the height of the Gulf War, I found in the *New York Times* this quote from Loren Thompson, director of the national security program at Georgetown University, explaining why the Pentagon wasn't releasing information about deaths in Iraq. When bomb damage is listed only in technical terms, he said, "you avoid talking about lives lost, and that serves both an esthetic and a practical purpose."

The esthetic and practical purpose, of course, is the loss of empathy. We seem to be living in the age of anesthesia, and it's no wonder. Confronted with knowledge of dozens of apparently random disasters each day, what can a human heart do but slam its doors? No mortal can grieve that much. We didn't evolve to cope with tragedy on a global scale. Our defense is to pretend there's no thread of event that connects us, and that those lives are somehow not precious and real like our own. It's a practical strategy, to some ends, but the loss of empathy is also the loss of humanity, and that's no small tradeoff.

Art is the antidote that can call us back from the edge of numbness, restoring the ability to feel for another. By virtue of that power, it is political, regardless of content. If *Jane Eyre* is a great romance, it has also given thousands of men a female experience, and a chance to feel the constraints that weighed upon women of Jane's time. Through art, a woman can give a male reader the unparalleled athletic accomplishment of childbirth, or the annihilation of being raped; if every man knew both those things, I would expect the world to change tomorrow. We have all heard plenty about each other's troubles, but evidently it's not enough to be told, it has to be lived. And art is so very nearly the same as life.

I *know*, for example, that slavery was heinous, but the fate of sixty million slaves is too big a thing for a heart to understand. So it was not until I read Toni Morrison's *Beloved* that I honestly felt that truth. When Sethe killed her children rather than have them grow up in slavery, I was so far from my sheltered self I knew the horror that could make infanticide an act of love. Morrison carved the tragedy of those sixty million, to whom the book is dedicated, into something small and dense and real enough to fit through the door, get in my heart, and explode. This is how a novel can be more true than a newspaper.

One of my favorite writings about writing is this excerpt from Ursula K. Le Guin's introduction to her science-fiction novel *The Left Hand of Darkness*, in which she discusses fiction's role in what we call the truth:

Open your eyes; listen, listen. That is what the novelists say. But they don't tell you what you will see and hear. All they can tell you is what they have seen and heard, in their time in this world, a third of it spent in sleep and dreaming, another third of it spent in telling lies.

. . . Fiction writers, at least in their braver moments, do desire the truth: to know it, speak it, serve it. But they go about it in a peculiar and devious way, which consists in inventing persons, places, and events which never did and never will exist or occur, and telling about these fictions in detail and at length and with a great deal of emotion, and then when they are done writing down this pack of lies, they say, There! That's the truth!

. . . In reading a novel, any novel, we have to know perfectly well that the whole thing is nonsense, and then, while reading, believe every word of it. Finally, when we're done with it, we may find that we're a bit different from what we were before we read it, that we have been changed a little. . . crossed a street we never crossed before. But it's very hard to *say* just what we learned, how we were changed.

The artist deals with what cannot be said in words.

The artist whose medium is fiction does this *in words*. The novelist says in words what cannot be said in words.

This baffling manifesto is a command that rules my writing life. I believe it means there are truths we all know, but can't make ourselves feel: Slavery was horrible. Love thy neighbor as thyself, or we'll all go to hell in a handbasket. These are things that cannot be said in words because they're too familiar to move us, too big and bald and flat to penetrate our souls. The artist must craft missiles to deliver these truths so unerringly to the right place inside of us we are left panting, with no possibility of doubting they are true. The novelist must do this in story, image, and character. And make the reader believe.

To speak of this process as something that must fall either into the camp of "political" or "pure" is frankly absurd. Good art is political, whether it means to be or not, insofar as it provides the chance to understand points of view alien to our own. Its nature is the opposite of spiritual meanness, bigotry, and warfare. If it is disturbing at times, or unpalatable, it may be a good idea to buy it anyway.

In time, I came back from political exile. Not with my tail between my legs, having discovered the U.S.A. was after all the greatest place in the world. On the contrary, I loved the new experience of safety, the freedom to walk anywhere I pleased at any time of day, and the connected moral comfort of a society that cares for all its children, provides universal health care, and allows no one to be destitute. All these foreign things, and more, I loved: the sound of the ocean in my window, and the towering poinsettia trees that blossomed along the roadsides from Christmas till Easter. I missed a few things: Mexican food, certain familiar music on the radio, the blush of a Tucson sunset running hot and sweet up the face of the Santa Catalina Mountains. And I missed the sound of my mother tongue. By accident, it turns out, I've been apprenticed as a writer to my own language and culture. In the midst of a deeply American novel, high and dry in the Canary Isles, I had to beg friends back home for mundanities I couldn't recall—figures of speech, car makes, even commercial jingles.

More than anything, though, I missed people, the beloved relatives and friends I left behind. I had new friends, but it was finally on account of the old ones that I prepared to give up the expatriate's life.

As the time drew near, my feet balked. I dreaded leaving my kind new place to return to the land of the free (*free* to live behind locks at all times; *free* to walk in the evenings from library to parked car with sheer terror in my heart) and the home of the brave (well, yes, *brave*). The land where 7 percent of the world's souls guzzle the lion's share of the world's goods, pitch out a yearly average of sixteen hundred pounds of garbage apiece, and still can drive past homeless neighbors with little awareness of wrongdoing or alternatives. The place I was told to love or leave.

I found I could do neither. Not wholeheartedly. But like the boy who fought the Jabberwock in *Through the Looking Glass*, I took my vorpal sword in hand. For the sake of people who love me and the sight of mountains that move my soul, I would come galumphing back, to face the tyranny of words without meaning and monsters beyond my ken.

I came back because leaving was selfish. A country can be flawed as a marriage or a family or a person is flawed, but "Love it or leave it" is a coward's slogan. There's more honor in "Love it and get it right." Love it, love it. Love it and never shut up.

THE FOREST IN THE SEEDS

In the springtime of my twenty-fifth year, and my first as a graduate student in ecology, I was seriously introduced to biological field research. The project to which I was assigned involved sitting in a mesquite thicket in the southern Arizona sun, watching a species of territorial lizard do, quite frankly, almost nothing. For hours and hours, day after day. It was stultifying. When I'd signed on as a rookie animal behaviorist, I suppose I was thinking of Konrad Lorenz's curiously malimprinted geese, who thought he was Mama Goose and followed him around; or of legendary Iwo, the genius macaque, who invented grain winnowing and introduced it to her tribe. Visions of sandhill cranes danced in my head. And here I had washed up instead in the land of torpid lizards. I could only be grateful that my subjects at least had *heartbeats*, and pity my botanically inclined colleagues who were counting pollen grains under a microscope, or literally watching the grass grow.

Nature does not move in mysterious ways, really. She just moves so slowly we're inclined to lose patience and stop watching before she gets around to the revelations. The natural historians of the nineteenth century knew this, or at any rate they had no reason to expect bells and whistles, and they had the luxury of writing for an audience with an attention span. Charles Darwin charmingly suggests as much in his introduction to *On the Origin of Species*: "It occurred to me, in 1837, that something might perhaps be made out of this question [of the origin of species] by patiently accumulating and reflecting on all sorts of facts which could possibly have any bearing on it." Twenty-two years later he'd reflected on everything from slave-making ants to the Greenland whale and set it all down on paper, and for any reader willing to spend a portion of a lifetime with it, it remains a thorough masterpiece.

Henry David Thoreau, Darwin's contemporary, shared the penchant for accumulation and reflection, and while he did not shake the scientific paradigm so profoundly, he brought to his work an expansive poetic sensibility. Like other modern fans of his who had long since finished all the Thoreau in print, I rejoiced when Bradley P. Dean compiled from the massive notebooks of Thoreau's last two years a collection of previously unpublished writings, *Faith in a Seed*. The book contains fragmentary treatises on wild fruits, weeds and grasses, and the succession of forest trees. But the centerpiece is Thoreau's last important manuscript, *The Dispersion of Seeds*, in which he meticulously noted methods of seed ripening and dispersal, germination, and growth of a great many species: pines, willows, cherries, milkweeds, eight kinds of tick clover, and virtually every other plant known to the neighborhood of Concord, Massachusetts. With a categorical thoroughness akin to Darwin's, Thoreau intended to prove his conviction—which was still in dispute at the time—that new plants do not spontaneously generate but, rather, grow always and only from seeds.

It's hard to imagine grown men of science being uncertain of a thing that our first-graders now might snub as a science-fair project. ("A bean in a Dixie cup? That's *kid stuff*," mine once hooted.) So the energy Thoreau brings to this argument may seem quaint for its obsolescence. But there is something wonderful to be gained from a two-hundred-page walk through the woods with a scientist from a century and a quarter ago. Thoreau had just read *On the Origin of Species*, and was clearly moving away from the travelogue format of his "excursion" writings, toward an articulation of unifying principles; he was attempting to see the forest among his trees. In his observations of plant communities he touched on succession, allelopathy, and other concepts that would not have names until the birth of the science of ecology in the next century.

His gifts as a writer, though, transcended his contributions to natural science. Thoreau dismissed the notion that poetry and science are incompatible, and captured for his readers the simple wonder we hastily leave behind in the age of reason. "How impatient, how rampant, how precocious these osiers," he wrote of the willows along his pond. "Some derive their Latin name *Salix* from *salire*, 'to leap,' they spring up so rapidly—they are so salient. They have hardly made two shoots from the sand in as many springs, when silvery catkins burst out along them, and anon golden blossoms and downy seeds, spreading their race with incredible rapidity."

He admired the trees for their ingenuity, and praised the wind that catches their seeds for its unfailing providence. He carefully watched the ways and means of the seed-scattering creatures: squirrels, foxes, birds (including, nostalgically, the now extinct ivory-billed woodpecker), a wading moose or cow, or "a wading pickerel fisher of the old school, who does not mind if his clothes be wet," and even little boys who blow the seeds off dandelion heads to find out whether their mothers want them. ("If they blow off all the seeds at one puff, which they rarely do, then they are not wanted.")

As I made my leisurely way through Thoreau's final book I found myself turning down the corner of nearly every other page to note an arresting moment of prose; eventually I realized I was admiring not specific bits of information but the man himself. As a Transcendentalist, Thoreau understood that the scientist and the science are inseparable, and he insinuated himself into his observations in a way that modern science writers, we virtuosos of the passive voice, have been trained carefully to forsake.

"I went forth on the afternoon of October 17th," one section begins, "expressly to ascertain how chestnuts are propagated." American chestnuts are now as dead as the ivory-billed woodpeckers, but still a reader can watch this bearded, wide-eyed man— who would within two years of that journal entry be dead himself—inhaling an autumn day and focusing his powers not only on the chestnuts but also on his own heart and the folkloric tenor of his village. "A squirrel goes a-chestnutting perhaps as far as the boys do, and when he gets there he does not have to shake or club the tree, or wait for frost to open the burrs, but he walks up to the burrs and cuts them off and strews the ground with them before they have opened. . . . The jays scream and the red squirrels scold while you are clubbing and shaking the chestnut trees, for they are there on the same errand, and two of a trade never agree."

Another passage exclaims, "Consider what a vast work these forest planters are doing! So far as our noblest hardwood forests are concerned, the animals, especially squirrels and jays, are our greatest and almost only benefactors.

"But what is the character of our gratitude to these squirrels? . . . Are they on our pension list? Have we in any way recognized their services? . . . We should be more civilized as well as humane if we recognized once a year by some symbolical ceremony the part which the squirrel plays in the economy of Nature."

Faith in a Seed is infused with Thoreau's delight, his meticulous curiosity and his inspiring patience. Across the silence of 125 years, during which an unforeseeable glut of hurry has descended, he exhorts us to slow down and take notice, to learn how to watch seeds become trees. This is the kind of book that should be forced on students, probably against their will. When I recall my lizard-watching days I can sympathize with their restlessness, but I also long for all of us to rescue ourselves from the tyranny of impatience. Like cartoon characters, we seem to be running full tilt through the air beyond the edge of the cliff with our minds on something else. In *Earth in the Balance,* Al Gore poignantly discusses this detached relationship between humans and our earth. He reports that in a 1991 poll that asked the American people for their views about the role we should play in the world, an incredible 93 percent supported a proposal for "the U.S. using its position to get other countries to join together to take action against world environmental problems." And yet at the same time, he writes, "Almost every poll shows Americans decisively rejecting higher taxes on fossil fuels, even though that proposal is one of the logical first steps in changing our policies in a manner consistent with a more responsible approach to the environment." Is it possible we just couldn't sit still long enough to make the connection?

Recently, as I gave a lecture to a college class on writing and environmental activism, a student asked me, "Why can't we just teach people about this stuff in TV commercials?" The question was both naïve and astute. As a nation we will never defer to the endangered spotted owl (let alone declare a National Squirrel Holiday, as Thoreau suggested) until we are much more widely educated. But the things we will have to know—concepts of food chain, habitat, selection pressure and adaptation, and the ways all species depend on others—are complex ideas that just won't fit into a thirty-second spot. Evolution can't be explained in a sound bite.

Even well-intentioned educational endeavors like carefully edited nature films, and the easy access to exotic animals offered by zoos, are tailored to our impatience. They lead us to expect nature will be all storm and no lull. It's a dangerous habit. Natural-history writer Robert Michael Pyle asks: "If we can watch rhinos mating in our living rooms, who's going to notice the wren in the back yard?"

The real Wild Kingdom is as small and brown as a wren, as tedious as a squirrel turning back the scales of a pine cone to capture its seeds, as quiet as a milkweed seed on the wind—the long, slow stillness between takes. This, I think, is the message in the bottle from Thoreau, the man who noticed a clump of seeds caught in the end of a cow's whisking tail and wondered enviously what finds were presenting themselves to the laborers picking wool in nearby factories. "I do not see," he wrote, "but the seeds which are ripened in New England may plant themselves in Pennsylvania. At any rate, I am interested in the fate or success of every such venture which the autumn sends forth."

What a life it must have been, to seize time for this much wonder. If only we could recover faith in a seed—and in all the other complicated marvels that can't fit in a sound bite. Then we humans might truly know the glory of knowing our place.

CAREFUL WHAT YOU LET IN THE DOOR

Once in a while I've heard people in my profession claim, with the back of a hand thrown across their foreheads, that it's a curse to be a writer. I am inclined to tell them: Get real. It's a curse to be one of those people who have to put asphalt on the highway with what looks like the back of a janitor's broom in the middle of July. I've never done that, and I'm deeply happy about it. But I have held about twenty jobs in my life that I might call a curse, including baby-sitting a pair of twins named Aristotle and Alexander, who had the energy and will of spider monkeys and a language of their own invention; also, scrubbing toilets for people who spoke of me as The Cleaning Lady. (I was barely twenty years old; in no other setting did I get called, at that time, a lady.) If there's no statute of limitations on this list, I'll even mention picking tent caterpillars off my Dad's apple trees for the salary of a penny apiece. (Caterpillar disposal, involving gasoline, was included in the price.)

Writing is no curse. The writing life has incomparable advantages: flexible hours, mental challenge, the wardrobe—you can go to work in bunny slippers if you want to. The money, well, that is sometimes a snag, but if you keep your nose to the grindstone the benefits accrue. You can support yourself. And in time, if you're truly blessed, you'll begin to get *mail*. You'll bring it home by the carload, tear it open, and find out everything you've ever done right in this world, and wrong. The mail will bring you more applause and brickbats and requests and advice and small, perfect bouquets than you can ever answer or even acknowledge. Its presence will cheer you on gloomy days, and guide you through the straits of your own conscience. It will stand as proof that you're blessed.

I have received, entirely unsolicited: advice on dog racing ("conventional wisdom has it that the outside post positions are bad and—over the long haul—more low numbers come in than high") and natural pest control ("I have never had success combatting flea beetles with diatomaceous earth"); information on how to order foam clothing; a Christmas card from the Dan Quayle family; and outlines for approximately ten thousand novels based on other people's relatives' lives. I've received works of art that I adored, many of which are hanging on my walls. After publishing a novel called *Pigs in Heaven*, I received via U.S. mail more pig-oriented items than you might have imagined to exist. (I'm pretty sure I'm going to call my next novel *Mustang Convertible Dreams*.)

I've received this information on how to live forever: "I suggest a petition to Masauwu, Spirit of Death, Owner of Fire and Master of the Upper World. Sanction may be gained to the sipapuni for shelter during the destruction of the Fourth World and re-emergence to the Fifth. Even if it doesn't work, it's worth a shot."

Also this useful tip: "Dear Barbara Kingsolver, It appears to me that your last name is to be derived from *Gundisalv*, a name compounded by the Visigoths of Northwestern Spain from the Old Germanic elements *gundi*, meaning 'battle,' and *alf*, meaning 'elf.'"

(When I passed this on to my relatives, they started calling me the old Battle-Elf.)

A New York City reader wrote: "Dear Ms. Kingsolver, Your novels have to be the most implausible, coincident ridden, knee jerking exhibits of liberalism and corny sentimentalism that I have ever read. P.S. I like them pretty well."

And a befuddled fan in California wrote: ". . . I am very interested in animal consciousness, as well as dreams, and I bought your book *Animal Dreams* because I believed it to be a book I had heard about on the radio once, called (as I am now aware) *Animal Dreaming*. When I sat down and saw it was fiction and that I had paid $20 for it, I thought: Mistake!"

I haven't found a use for this information: "Dear Ms. Kingsolver, I am 23 years old, have 3 tattoos, and 2 college degrees that are doing me no good."

This one was slightly more upbeat: "I lent my library copy of *The Bean Trees* to a friend who normally hates everything (seriously, she's very depressing). She loved it! That is, until it was stolen from her car. We had to pay the $16 replacement cost plus library fine."

There is a type of letter that comes from remarkable adolescent girls, like this one: "Dear Mrs. Kingsolver, I wrote you before that I was writing a novel and you encouraged me to do so. I finished it. It's called *The Little Cabin in the Woods*. Then I wrote *The*

Dark Crystal, followed by *Sky Eyes*, and *Fireball in the Night, The Clue, Blue Dawn, The Princess Bride,* and *Emily*, which is a hypothetical look at what might happen to me if my parents suddenly died."

There are also ever so many assorted requests from people who would like you to do them some small favor. For example:

Dear Ms. Kingsolver, Enclosed is something I've written. I'd appreciate it if you could get Harper & Row to publish it. I suggest it be marketed as an Inspirational Essay.

Dear Ms. Kingsolver, Our book club would appreciate my sharing any materials from you. Would you send me:

Photos
Interviews/Statements
Biographical Data
Your comments on the book
Reviews
Career Plans/Goals [Apparently they are still expecting me to do something productive. . . .]

Dear Mrs. Kingsolver, I am doing a paper for school, on why you should be considered a great American author. In this paper I must classify your writing as following an American tradition: Puritanism, Romanticism, Trancendentalism, Rationalism, Idealism, and Realism. I also must prove that you contribute something to American Literature. . . . I would greatly appreciate having your opinion on this matter and any suggestions you might have. My paper is due in two weeks.

Best, of course, are the letters that go straight to your head, like this one:

> Dear Barbara, I just finished reading *The Bean Trees* for
> the fourth time since I bought it through a book club.
> Please, please, please write more books!

I walked on air for days, imagining someone actually reading my book four times, scanning it for every alliteration and metaphor I'd buried in its pages. Then I considered the return address: South Padre Island, Texas. I've been to South Padre Island, Texas, and so I know. If you lived there, you would have no choice but to read whatever washed up on shore, or otherwise fell into your hands, four times at a dead minimum. My hunch bore out a few years later when I heard from the correspondent again:

> Dear Barbara, I wrote you in 1988 to express admiration
> for your novel, *The Bean Trees*. Since then I understand
> you have two more books out. . . . Things move slowly
> in South Texas. The bookstore filed Chapter 7 two
> months ago. In two years they managed to get me a copy
> of *Holding the Line* by Dwight D. Eisenhower (it was
> soporific). . . . I will send money order, personal cheque,
> bank card, jewels, or whatever is necessary. I'll eat sand. [I
> immediately sent copies of everything I'd ever written.]

I'm grateful beyond words for reader mail, which keeps me going through the days when I can't believe in myself, or literature in general. That is the blessing. And perhaps it's also the curse, if the writing life is cursed, because readers tug on the writer's solitude and complacency. One of the few pieces of advice I ever give other writers, if they ask for it, is to try to write with no one looking over your shoulder. It's heaven, if you can do it. But inevitably they come, those ghosts and battle-elves peeking in through the study door left ajar, and even if they are not allowed a vote, they force the writer to answer all the disparate voices rattling inside her own psyche. The compliments must be accepted, and so, too, the thoughtful complaints. Once in a while a letter rocks my foundations, causing me to question once again the things I thought I knew about art and responsibility.

This was one of those:

I've decided you might like to hear [she wrote] about one woman's response to *Animal Dreams*. I sailed through the book till I got to Hallie's kidnapping. Then I stopped cold and skimmed ahead, reading only with my head, keeping my heart out of it, because I began to realize that if she was going to get killed I didn't want to read the rest.

Like many women, and men, in America, I was abused as a child, and when I started censoring TV for my own small children, I decided to stop watching violent TV shows myself. It really made life better. . . . Yes, there is violence all around us. I read the news and even sometimes watch it on TV. But that's real. To invent violence that didn't really happen, even for the noblest of motives, like making everybody see how stupid war is, also puts it out there as entertainment. On a certain level, even people who are moved by the nobility and poignance of it all are also going to get off on it in a

way that is absolutely counterproductive to the end of
ending violence. . . .

I replied to this letter with a brief, inadequate response, and I
haven't stopped thinking about it since. Oddly, in the same week
I got another letter addressing the violence in *Animal Dreams*
from a different perspective, from a Sister of St. Agnes, in
Milwaukee:

> I am writing to thank you. I picked up *Animal Dreams*
> because I was eager to read any book dedicated to Ben
> Linder and daring to hold up a mirror to the horrible
> devastation our country has visited upon
> Nicaragua. . . . All through the eighties, Reagan's policy
> was driving me nuts. . . . Then in early 1990 it hit
> home. It was then we got word that two of our sisters
> were ambushed on a lonely road in Nicaragua. Killed
> by U.S.-supplied armaments. One was a North
> American, a Milwaukee native, and the other was a
> Miskito Indian woman who had been in vows less than
> a year. . . . I want to thank you for your novel, which
> says something hopeful about death and the life that
> can come from death.

The sentiments in the second letter don't change the signifi-
cance of the first. I can't in good conscience ignore either one. I
don't know whether my convictions about art—and particularly,
art that contains violence—will ever be allowed to settle into a
comfortable position. They have been revising themselves for a
long, long time, roaming restlessly over the options, continually
exhorted by the ghosts that bless and curse.

As an adolescent girl, I had a secret yellow notebook I filled with stories. They were written in a crabbed cursive, set mostly in places I had never been, like Mexico and the Andes, and the protagonists of these stories were always boys. What's more, they were almost always maimed in some way. One of my heroes, I remember, had been blinded, and yet he still managed to canoe across a lake and climb a mountain. Another one had a clubfoot, and he won a scholarship to leave his small folkloric village and study art. When I was eleven, I'm sure I didn't know what a clubfoot was; I think I had some vague idea that if someone clubbed you on the foot, then you would have a clubfoot.

I was very much like that girl who has written *The Princess Bride* and *The Dark Crystal* and thirty-five other novels and is now wondering how the plot possibilities will open up if she knocks off her parents. When I was her age, I wasn't remotely conscious of what it took to make good writing. I was just looking for drama and impact, and the only way I could see to get that onto a page was to write about events that, if they happened to you in real life, would tend to make a big impact.

I didn't realize that it's *emotion*, not *event*, that creates a dynamic response in the mind of a reader. The artist's job is to sink a taproot in the reader's brain that will grow downward and find a path into the reader's soul and experience, so that some new emotional inflorescence will grow out of it.

Of course, the writer has to do this for many readers at a time, without ever having met any of them, knowing nothing about them except that they're human and have mostly all lived on the same earth. So it's a challenge. Lacking the skills to pull that off, it's common for beginning writers to fall back on the put-out-his-eyes-and-make-him-climb-a-mountain tract. Some years ago as a judge in a fiction contest, I read the unscreened entries of a few hundred aspiring writers and, I swear, three out of four contained unfortunate wretches trapped in wheelchairs in burning buildings. *That* job was a curse.

In time, with practice, you learn that violence isn't a necessary component of exciting art. You can substitute metaphor and imagery for the clubfoot. And then comes the question: If you don't *have* to, why would you *want* to create violence in art? Are there any good reasons? Maybe yes. Maybe no.

To some extent I agree with my correspondent who wrote that inventing violence, even for the noblest of motives, might necessarily be promoting violence as entertainment. The equation of fun-for-pay with the infliction of pain makes me very uneasy. Very often it's done with a cast of morality thrown over the whole thing, as though that might redeem it—for example, in the genre I call Slice & Dice movies, to which teenagers flock in droves. For an hour and a half you get to see attractive, terrified young women and a good deal of spurting blood; then the colorful criminal is apprehended and we get to see *his* spurting blood; so justice was served. It wasn't really okay that he was going around damaging people with farm implements, so it's not really condoning violence. But then, I wonder, why did we have to watch? And more to the point, why did we *pay* to watch, enabling legions of grown-ups to earn their living fabricating the realistic illusion of terrified young women spurting blood?

Sometimes the same formula is passed off as something more noble, because of higher production values and more imaginative criminals. The film *Silence of the Lambs* was one of the great critical successes of our time, and for that reason I felt obliged to see it, even though I hate feeling sick with fear and suspense, and have never understood why I should pay for that sensation when it's easy enough to come by it for free. But I watched, on a friend's VCR; got up and left the room every time somebody's flesh was in danger, which was most of the movie; and afterward felt ripped off. It turns out, I'd rented the convincing illusion of helpless, attractive women being jeopardized, tortured, or dead, for no good reason I could think of after it was over. You may disagree. Obviously most people in the world do. But I'm uncomfortable with the huge popularity of that film. I know, now, I should have stuck with my instincts and skipped it. I felt the way many African Americans probably felt watching the old Star Trek plots in which, any time you saw an anonymous lieutenant in an Afro beaming down to Planet X with the landing party of white guys, you knew somebody was going to bite the dust on Planet X, and you knew who it was going to be. Anyone who complained about that kind of story line, at the time, probably would have seemed overly sensitive. When nobody else can see what's driving you crazy, it's easy to feel you're making it up. Even when you're not.

When I watch a film whose plot capitalizes on the vulnerability of women to torturers, maimers, rapists, and maniacs, I take it personally. I feel preyed upon. I don't enjoy sitting through another woman's misery, even if I keep telling myself that her big problems there are really all just ketchup. It still hurts to watch. For me, a recreation of simple violence has no recreational value. So why would I ever create an act of violence in a novel?

My answer has to do with the fact that I don't consider a novel to be a purely recreational vehicle. I think of it as an outlet for my despair, my delight, my considered opinions, and all the things that strike me as absolute and essential, worked out in words. When I wrote in my secret yellow notebook, it was not for other people, and I still write for mostly the same private reasons. It's my principal way of becoming reassured I'm still alive: I have come through this many of my allotted days, watched the passing of life on earth, made something of it and nailed it to the page. Having written, I find I'm often willing to send it on, in case someone else also needs this kind of reassurance. Art is entertainment but it's also celebration, condolence, exploration, duty, and communion. The artistic consummation of a novel is created by the author and reader together, in an act of joint imagination, and that's not to be taken lightly.

One of the extremely valuable things to be done with the power of fiction is the connection of events with their consequences. And violence, above all else, is a thing with consequences. The difference between the violence in great novels like *War and Peace* or *Beloved*, and the contents of a Slice & Dice movie (or a Slice & Dice *book*; there are plenty of those) is the matter of context. Occasionally I make the error of seeing an adventure movie that I've been assured isn't violent, and inevitably, throughout the movie people are dying like flies. But like flies they don't have personalities, they are just *there*. They fall off of things or they get shot and they are gone, like the unfortunate lieutenant of color on Planet X. We never knew the guy so we don't feel a thing, and we don't have to sit through the funeral. If you had to sit through all the funerals, most TV shows would be seven hours long. But you don't.

See enough of this bang-you're-dead kind of thing and you'll start to go numb around the edges, I guarantee. On some level you will start to believe that a violent act has no consequences. Researchers in social psychology have known for decades that watching violence makes a person more likely to participate in violence. Many people in the entertainment industry would have us believe otherwise, and so these studies are controversial, but they are mostly unequivocal. A review article written in 1991 by Wendy Wood, Frank Wong, and Gregory Chachere examined the body of research in this field, conducted in both laboratory and natural social settings, and they found that exposure to media violence significantly enhanced viewers' aggressive behavior. Hundreds of other psychologists stand in agreement. They suggest many different mechanisms for the causal link between watching and doing: increased physiological arousal; decreased inhibitions; instrumental learning and modeling of aggressive acts; and decreased sensitivity toward violent acts. It boils down to one thing: we learn about the world through our senses, like any other creature. Watch your mother make a hundred tortillas, and you know how to love, live with, and manufacture a tortilla. Watch a hundred violent deaths and that, too, is your familiar. That the deaths were all faked is apparently incidental to the hardware in our heads that brings us learning. A trick on the eye works a trick on the psyche as well, for although our brains know it is only ketchup, in our animal soul it registers as blood. Blood without consequence.

So it happened, one day in Florida, that a thirteen-year-old shot a man in the head because he took two slices of pizza when he was only offered one. It has happened a thousand times over, will happen again tomorrow, and I hardly wonder why. That child believed the scene would fade out after he shot the gun, and then the world would be new again.

The simple, intense exposure of a vicious act, in film or in literature, is entirely different from a story that includes both the violence and its painful consequences. I can't even call these two things by the same name. Those who like to say there is nothing new under the sun will claim that TV is no more violent than

Shakespeare. But three average nights of prime-time TV contain as many acts of violence as all thirty-seven of Shakespeare's plays put together end to end—and quantity is only partly the point. More importantly, there is also a world of difference in the context. Think of all we learn of the world from poor Hamlet: the whole play is a chain of terrible consequences that fall one after another from the murder of his father. It's about bereavement, guilt, and unbearable loss. Hamlet "raised a sigh so piteous and profound as it did seem to shatter all his bulk and end his being." That is a tragedy that has earned its place.

I find I'm prepared to commit an act of violence in the written word if, and only if, it meets two criteria: first, the act must be embedded in the story of its consequences. Second, the fictional violence must be connected with the authentic world. It matters to me, for example, that we citizens of the U.S. bought guns and dressed up an army that killed plain, earnest people in Nicaragua who were trying only to find peace and a kinder way of life. I wanted to bring that evil piece of history into a story, in a way that would make a reader feel sadness and dread but still keep reading, becoming convinced it was necessary to care. So I invented Hallie Noline, and caused her to die. I did it because this happened, not to imaginary Hallie but to thousands of real people. One of them was a hydroelectric engineer in his twenties from Portland, Oregon, named Ben Linder, whose family I dearly love, and whose death is permanently grieved; *Animal Dreams* is dedicated to his memory. I would write that story again, because people forget, and I want us to remember.

I'm sure *Silence of the Lambs* had its reasons, too. Possibly its creators, who are a vastly talented lot, were trying to evoke in us a hatred of psycho-killers. But I should have exercised my right to stay away, on the grounds that I was already pretty clear about being no friend to psycho-killers. And the woman who wrote to tell me she closed the book on Hallie's death already knew enough, too. She did the right thing.

I will not argue for censorship, except from the grassroots up: my argument is for making choices about what we consume. The artist is blessed and cursed with a kind of power, but so are the reader and viewer. The story no longer belongs to the author once it's come to live in your head. By then, it's part of your life. So be careful what you let in the door, is my advice. It should not make you feel numb, or bored, or demeaned, or less than human. But I think it's all right if it makes you cry some, or feel understood, or long to eat sand for want of more, or even change your life a little. It's a story. That's what happens.

THE NOT-SO-DEADLY SIN

Write a nonfiction book, and be prepared for the legion of readers who are going to doubt your facts. But write a novel, and get ready for the world to assume every word is true.

Whenever I am queried about my fiction, if people want to know something in particular they nearly always want to know the same thing: How much is autobiographical? Did it all really happen, in exactly that way? Was my childhood like that? Which character is me? Commonly people don't ask, they just assume. I get letters of sympathy for the loss of my sister (the heroine of one of my novels lost her sister) and my father (ditto, same novel). Since one of my characters adopted a Cherokee child, I get advice about cross-cultural adoptions. And so on.

My sister and parents are alive and well, thanks. I don't have an adopted child. The mute waif named Turtle who appears in two of my novels is the polar opposite of my own Camille—a sunny, blonde child who spoke her first word at eight months and hasn't stopped talking since. At the time I invented Turtle, I had no child at all. Mine came later, and I didn't find her in a car, as happened in *The Bean Trees*. Mine was harder to produce. I never use my own family and friends as the basis of fictional characters, mainly because I would like them to remain my family and friends. And secondarily, because I believe the purpose of art is not to photocopy life but distill it, learn from it, improve on it, embroider tiny disjunct pieces of it into something insightful and entirely new. As Marc Chagall said, "Great art picks up where nature ends."

I know, in real life, many fascinating people; every one of them has limits on what she or he can be talked into. Most, in fact, will ask for my recommendations on their love lives or vacation plans, then reliably do the opposite. When I'm writing a story, I can't mess around with that kind of free spirit. I need characters I can count on to do what I say—take on a foundling baby rather than call the police; fall in love with my self-effacing heroine rather than the sturdy, good-looking divorcée down the street; pursue a passion for cockfighting, then give it all up at a lover's request; die for honor; own up to guilt. What's more, they must do it all *convincingly*. That means they have to be carrying in their psyches all the right motives—the exact combination of past experiences that will lead them to their appointment with my contrived epiphany. Trying to graft a plot onto the real-life history of anyone I actually know, including myself, would be as fruitless as lashing a citrus branch onto the trunk of an apple tree. It would look improbable. It would wither and die. Better to plant a seed in the good dirt of imagination. Grow a whole story from scratch.

Most people readily acknowledge the difference between life and art. Why, then, do so many artists keep answering the same question again and again? No, none of those characters is me. It's not my life, I made it up. Yes, *all of it!* Strangers' assumptions about deaths in my family and the like, odd as this might seem, have caused us some genuine pain. How I wish my art could stand apart from us, carrying no more suggestions about my private life than the work of, say, a stonemason or a tree surgeon. I was raised to be polite, but sometimes I'm inclined to get cranky and bark about this: Give us writers a little credit, will you? We're not just keeping a diary here, we're inventing! Why can't you believe we're capable of making up a story from scratch? Of stringing together a long, elaborate *lie*, for heaven's sake?

When it's put that way, it dawns on me that this may be the snag—the part about lying. In the book-jacket photos I look like a decent girl, and decent girls don't lie. That social axiom runs deep, possibly deeper than any other. The first important moral value we teach our children, after "don't hit your sister," is the difference between fantasy and truth. Trying to pass off one for the other is a punishable infraction, and a lesson that sticks for life. Whether or not we are perfectly honest in adulthood, we *should* be, and we know that on a visceral level. So visceral, in fact, a machine measuring heart rate and palm perspiration can fairly reliably detect a lie. We don't even have to think about it. Our hearts know.

So I suppose I should be relieved when people presume my stories are built around a wholesome veracity. They're saying, in effect, "You don't *look* like a sociopath." And it's true, I'm not; I pay my taxes and don't litter. Track down any grade-school teacher who knew me in childhood and she'll swear I was a goodie two-shoes even back then.

But ask my mother, and she'll tell you I always had a little trouble with the boundaries of truth. As the aerospace engineers say, I pushed the envelope. As a small child I gave my family regular updates on the white horse wearing a hat that lived in the closet. When I was slightly older, family vacations offered me the delightful opportunity to hang out alone in campground restrooms, intimating to strangers that I came from a foreign country and didn't comprehend English, or plumbing. When I got old enough to use public transportation by myself, my sport was to entertain other passengers with melodramatic personal histories that occurred to me on the spot. I was a nineteen-year-old cello virtuoso running away from my dreadful seventy-year-old husband; or I had a brain tumor, and was determined to see every state in the union by Greyhound in the remaining two months of my life; or I was a French anthropologist working with a team that had just uncovered the real cradle of human origins in a surprising but as-yet-undisclosable location. Oh, how my fellow passengers' eyes would light up. People two rows ahead of me would put down their paperbacks, sling an elbow over the back of the seat, and ride all the rest of the way to Indianapolis backward, asking questions. I probably registered an increased heart rate and sweaty palms, but I couldn't stop myself. I strove for new heights in perjury, trying to see how absurd a yarn I could spin before some matron would finally frown at me over her specs and say, "Now really, dear."

No one ever did. I concluded that people want pretty desperately to be entertained, especially on long bus rides through flat midwestern cornfields.

For me, it was more than a pastime. It was the fulfillment of my own longing to reach through the fences that circumscribed my young life, and taste other pastures. Through my tales I dis-

covered not exactly myself but all the selves I might have been—
the ones I feared, the ones I hoped for, and the ones I'd never
know. None of them was me. Each of them was a beckoning path
into the woods of what might have been.

Eventually I found a socially acceptable outlet for my depravity. Now I spend hours each day, year after year, sitting at my desk
with a wicked smirk on my face, making up whopping, four-
hundred-page lies. Oh, what a life.

I do want to state for the record that I no longer have any
inclination toward real dishonesty; I don't bear false witness to
strangers or to friends. And I check my facts obsessively when
serving the journalist's or essayist's trade. So my mother isn't to
blame—she did, evidently, teach me to know true from false. I
gather I was just born with an excess of story, the way another
poor child might come into this world with extra fingers on each
hand. My imagination had more figment in it than my life could
contain, so some of it leaked out here and there. As I've matured,
I've learned to control the damage.

I don't believe I'm extraordinary on this account. Every one
of us, I think, is born with an excess of story. Listen quietly to a
group of toddlers at play: the lies will swarm around their heads,
thick as a tribe of bright butterflies, flickering gracefully from
one child to another, until they notice a grown-up has come
into the room—and in a sudden rush of wings the lies will vanish into air.

A little bit sad, isn't it? If you look it up, you'll find lying was
never registered as one of the seven deadly sins. (Pride—an ane-
mic sin if you ask me—is on that list, and so is gluttony, and of all
things, sloth. But not lying.) Yet, in the age of evidence and reason, it has gotten such a very bad name. When so many smart,
lively people keep insisting to me that all my stories must be true,

I begin to suspect they can't quite get their minds around the notion of pure fabrication.

I want to tell them: Stop a minute, right where you are. Relax your shoulders, shake your head and spine like a dog shaking off cold water. Tell that imperious voice in your head to be still, then close your eyes, and tap the well. Find the lie you are longing to tell. It's in there. When you manage to wrestle that first one out, a whole flood may gush out behind it. Take them up in your hands, drink their clarity, write them down in a secret book. Tell them to your children behind the golden door of "Once upon a time." Choose one chair at your dinner table, give it to a different family member each night, and declare it "the liar's seat."

Or take a long bus trip through the cornfields. You may find a new career.

REPRISE

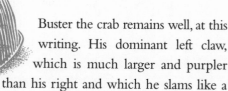

Buster the crab remains well, at this writing. His dominant left claw, which is much larger and purpler than his right and which he slams like a door behind him when he withdraws into his shell, is showing some wear. It's rumpled and split around the edges like an old laminated countertop. In fact, even though he has no greater adversary in his life than his own mood swings, he has recently managed to lose one of his antennae and is looking pretty dinged up. We think he may be preparing to molt. Crabs have this option: they can split themselves open from time to time and start life over with a fresh skin, complete with new appendages and even— if need be—whole regenerated eyes. The molting process itself is as astonishing as its results: the hermit preparing to shed its brittle skin will creep out of whatever seashell it's wearing at the moment, bury itself in damp sand, and inhale water (insofar as an animal with gills "inhales") until it has built up enough hydrostatic pressure to split its old casing and shuck it off. This is self-renewal at its fiercest and most tempting. It's the secret belief most of us carry forward from childhood, that we might have in us some-where the capacity, like Rumpelstiltskin, to rupture and transmo-grify out of a sheer tantrum of desire.

The crab's new skin is soft for a time, until it has the chance to dry and harden up like varnish. This is the brief period of its life when an edible marine crab becomes the potential delicacy known as soft-shelled crab. When the crab molts, it emerges larger; since its skin has no elasticity, this is the only way it can grow. If a newly molted hermit crab finds it can't fit back into the shell it parked nearby prior to molting, it may panic. My guide-book to hermit-crab care, written by Neal Pronek, advises that it's good to leave an assortment of shells of various sizes lying around just in case. Pronek waxes mostly pragmatic in his book, explaining for example that hermit crabs "will eat anything they find, from hard dog biscuits to a dead fish. . . where certain items are concerned, the deader the better," and also warning, "Don't expect an about-to-molt crab that loses a leg on Tuesday to pop up with a new one after a molt on Wednesday." But on the topic of hermit crabs stranded without shells, Pronek can hardly con-tain his alarm: "They'll start having nervous breakdowns. . . . They want those shells, and they'll do everything in their power to make sure that they don't get cut off from them. Pinch, scratch, smash, kill—whatever." Not something to mess around with. Since Buster started showing molting inclinations, we've sorted through our shell collections from every vacation in recent history (we knew we were saving these things for some good rea-son) and pulled out the cream of the crop. We believe we have got the situation in hand.

But one can't be sure. In the chapter called "Diseases and Ailments," Mr. Pronek offers darkly: "All of their ills boil down to the mysterious croak; the crab is outwardly well one day, dead the next." And so, while I can say that Buster remains well at this writing, around here we take nothing for granted.

After two days of gentle winter rains, the small pond behind my house is lapping at its banks, content as a well-fed kitten. This pond is a relative miracle. Several years ago I talked a man I knew who was handy with a bulldozer into damming up the narrow wash behind my house. This was not a creek by any stretch of imagination—even so thirsty an imagination as mine. It was only a little strait where, two or three times a year when the rain kept up for more than a day, water would run past in a hurry on its way to flood the road and drown out the odd passing Buick. All the rest of the time this little valley lay empty, a toasted rock patch pierced with cactus.

I cleared out the brush and, with what my bulldozer friend viewed as absurd optimism, directed the proceedings. After making a little hollow, we waterproofed the bottom and lined the sides with rocks, and then I could only stand by to see what would happen. When the rains came my pond filled. Its level rises and falls some, but for years now it has remained steadfastly *pond*, a small blue eye in the blistered face of desert.

That part was only hydrology and luck, no miracle. But this part is: within hours of its creation, my pond teemed with life. Backswimmers, whirligig beetles, and boatmen darted down through the watery strata. Water striders dimpled the surface. Tadpoles and water beetles rootled the furry bottom. Dragonflies hovered and delicately dipped their tails, laying eggs. Eggs hatched into creeping armadas of larvae. I can't imagine where all these creatures came from. There is no other permanent water for many miles around. How did they know? What jungle drums told them to come here? Surely there are not, as a matter of course, aquatic creatures dragging themselves by their elbows across the barren desert *just in case*?

I'm tempted to believe in spontaneous generation. Rushes have sprung up around the edges of my pond, coyotes and javelinas come down to drink and unabashedly wallow, nighthawks and little brown bats swoop down at night to snap insects out of the air. Mourning doves, smooth as cool gray stones, coo at their own reflections. Families of Gambel's quail come each and every spring morning, all lined up puffed and bustling with their seventeen children, Papa Quail in proud lead with his ridiculous black topknot feather boinging out ahead of him. Water lilies open their flowers at sunup and fold them, prim as praying hands, at dusk. A sleek male Cooper's hawk and a female great horned owl roost in the trees with their constant predators' eyes on dim-witted quail and vain dove, silently taking turns with the night and day shifts.

For several years that Cooper's hawk was the steadiest male presence in my life. I've stood alone in his shadow through many changes of season. I've been shattered and reassembled a few times over, and there have been long days when I felt my heart was simply somewhere else—possibly on ice, in one of those igloo coolers that show up in the news as they are carried importantly onto helicopters. "So what?" life asked, and went on whirling recklessly around me. Always, every minute, something is eating or being eaten, laying eggs, burrowing in mud, blooming, splitting its seams, dividing itself in two. What a messy marvel, fecundity.

That is how I became goddess of a small universe of my own

creation—more or less by accident. My subjects owe me their very lives. Blithely they ignore me. I stand on the banks, wide-eyed, receiving gifts in every season. In May the palo verde trees lean into their reflections, so heavy with blossoms the desert looks thick and deep with golden hoarfrost. In November the purple water lilies are struck numb with the first frost, continuing to try to open their final flowers in slow motion for the rest of the winter. Once, in August, I saw a tussle in the reeds that turned out to be two bull snakes making a meal of the same frog. Their dinner screeched piteously while the snakes' heads inched slowly closer together, each of them engulfing a drumstick, until there they were at last, nose to scaly nose. I watched with my knuckles in my mouth, anxious to see whether they would rip the frog in two like a pair of pants. As it turned out, they were nowhere near this civilized. They lunged and thrashed, their long bodies scrawling whole cursive alphabets into the rushes, until one of the snakes suddenly let go and curved away.

Last May, I saw a dragonfly as long as my hand—longer than an average-sized songbird. She circled and circled, flexing her body, trying to decide if my little lake was worthy of her precious eggs. She was almost absurdly colorful, sporting a bright green thorax and blue abdomen. Eventually she lit on the tip of the horsetail plant that sends long slender spikes up out of the water. She was joined on the tips of five adjacent stalks by five other dragonflies, all different: an orange-bodied one with orange wings, a yellow one, a blue-green one, one with a red head and purple tail, and a miniature one in zippy metallic blue. A dragon-fly bouquet. Be still, and the world is bound to turn herself inside out to entertain you. Everywhere you look, joyful noise is clanging to drown out quiet desperation. The choice is draw the blinds and shut it all out, or believe.

What to believe in, exactly, may never turn out to be half as important as the daring act of belief. A willingness to participate in sunlight, and the color red. An agreement to enter into a conspiracy with life, on behalf of both frog and snake, the predator and the prey, in order to come away changed.

The Cooper's hawk has been replaced as my significant other. A few weeks ago I was married, in the sight of pine-browed mountains, a forget-me-not sky, and nearly all the people I love most. This is not the end of the story; I know that much. With senseless mad joy, I'm undertaking what Samuel Johnson called the triumph of hope over experience—the second marriage.

Hope is an unbearably precious thing, worth its weight in feathers. If that's too much to think about, best to tuck it in a pocket anyway, and make it a habit. I was stomping through life in my seven-league boots, entirely unaware of how my life was about to snag on a doorframe, sending me staggering backward, on the day I met my future mate. But the *gris-gris* charm for luck in love, given me by a fetisher's apprentice in West Africa, was in its customary forgotten place, the watch pocket of my jeans. Now I keep it in a small clay jar among the potted plants in my bedroom window. Let the vandals carry off all but this—my hope.

On the day we met, my mate and I, he invited me to take a walk in the wooded hills of his farm in southwestern Virginia. I told him I loved the woods, and he took my word for that, and headed lickety-split up the mountainside. I ran after, tearing through blackberry briars with the options of getting hopelessly lost or keeping up.

He did remember, after all, that I was behind him. When he

reached the top of the mountain he waited, and we sat down together on a rock, listening to the stillness in the leaves. A song rang out through the branches, and because Steven is an ornithologist, he was able to tell me it was a rose-breasted grosbeak.

It sang again. He listened carefully, and said, "No, that's a scarlet tanager."

Either way, I was impressed by his ear for song. I asked him if he was sure. He said, "Yes, absolutely, that's a scarlet tanager."

And right then, exactly as he spoke, it came and landed on a branch directly in front of us, and it wasn't a scarlet tanager, it was a rose-breasted grosbeak.

Steven looked downcast; I shrugged and said, oh, what did it matter anyway. I think we both felt a little dismayed that this bird had come out of the woods to prove him wrong.

And then, directly in front of us, in a blaze of vermilion and perfect vindication, another bird landed—the one that had been singing, after all—and it was a scarlet tanager.

I had no idea this visitation of birds contained our future. Everything: risk, belief, forgiveness, being wrong, being right, finding how precariously similar those things are. And mainly, the whole possibility of bright red, singing marvels. What luck, I remember thinking. Here is a man who listens carefully to every voice.

He also had the patience to feed a wild fox who had whelped her pups in the pokeberry thicket behind the barn. Late that evening I sat on the stone porch steps of his old farmhouse and watched these two, man and fox, in their nightly ritual. He tossed out small scraps of meat, one after another; she approached, showing none of her hand but a pair of fierce green orbs in the dark—and accepted.

Eventually he would show the same patience in seeing me through my own wild fears and doubts, all the foul things my brain can turn over in a restless spell when it scrabbles around and around its cage at night. And so I have molted now, crawled out of my old empty banged-up skin with a fresh new life, and look here, what is this? I have regenerated a marriage, precious as a new eye.

I'm still feeling fairly soft-shelled. I'm too old to look at things the way I used to; too old, in fact, to look at anything closer than my own elbow without twinges of presbyopia (or, as one of my relatives calls it, "that Presbyterian thing"); I expect my next pair of glasses will need the extra window. So if I'm not quite the Bifocal Bride, I'm on the brink. I have a midlife vision of all things, including love and permanence. My dear mate and I will get to watch each other creak into old age and fall into uneasy truces with our own limbs—that's the *best* case, presuming we cleave together as we've hoped and promised. *It's a wonder anyone does this at all*, I think from time to time, as I'm visited by the specter of all I could lose.

When I was pregnant I felt like this too. People will claim that having children is a ticket to immortality, but in fact it merely doubles your stakes in mortality. You labor and you love and there you are, suddenly, with twice as many eyes in your house that could be put out, hearts that could be broken, new lives dearer than your own that could be taken from you. And still we do it, have children, right and left. We love and we lose, get hurled across the universe, put on a new shell, listen to the seasons.

Ah, the mysterious croak. Here today, gone tomorrow. It's the best reason I can think of to throw open the blinds and risk belief. Right now, this minute, time to move out into the grief and glory. High tide.

ACKNOWLEDGMENTS

"Creation Stories," in somewhat different form, was published as the introduction to *Southwest Stories*, eds. John Miller and Genevieve Morgan. San Francisco, Chronicle Books, 1993.

A brief portion of "Making Peace" appeared under that title in *Special Report,* November 1990.

"In Case You Ever Want to Go Home Again," is loosely based on an essay published in the *Lexington Herald-Leader,* September 16, 1990.

"How Mr. Dewey Decimal Saved My Life" is based on an address to the American Library Association Convention, New Orleans, June 1993.

"Life Without Go-Go Boots" appeared in Lands' End catalog, Spring 1990, and in the *Denver Post,* April 22, 1990.

"The Household Zen" appeared in different form as "A Clean Sweep," in the *New York Times Magazine,* December 30, 1990.

A much shorter version of "Semper Fi" was published under the title "Ah, Sweet Mystery of . . . Well, Not Exactly Love," in *Smithsonian,* June 1990.

"The Muscle Mystique" appeared as "After a Finger Workout, It's Great Pumping Iron," in *Smithsonian,* September 1990.

"Somebody's Baby" is loosely based on an essay entitled "Everybody's Somebody's Baby," published in the *New York Times Magazine,* February 9, 1992, and "License to Love," in *Parenting,* November 1994.

"Paradise Lost" appeared in different form as "Where the Map Stopped," in "The Sophisticated Traveler," the *New York Times Magazine,* May 17, 1992.

"Confessions of a Reluctant Rock Goddess" appeared in different form as a chapter in *Midlife Confidential: The Rock Bottom Remainders Tour America with Three Chords and an Attitude,* Dave Marsh, ed., Viking, 1994.

"Stone Soup" appeared in different form in *Parenting,* January 1995.

"The Spaces Between" is loosely based on an article entitled "Native American Culture Comes Alive in Phoenix," *Architectural Digest,* June 1993.

A brief portion of "Postcards from the Imaginary Mom" appeared in *I Should Have Stayed Home,* Roger Rapoport and Marguerita Castanera, eds., Book Passage Press, 1994.

"The Memory Place" appeared as a chapter in *Heart of the Land,* Joseph Barbato and Lisa Weinerman, eds., Pantheon, 1995.

"The Vibrations of Djoogbe" appeared in different form as "An Ancient Kingdom of Mystery and Magic" in "The Sophisticated Traveler," the *New York Times Magazine,* September 12, 1993.

ACKNOWLEDGMENTS

"Infernal Paradise" appeared in slightly different form as "Hawaii Preserved," in "The Sophisticated Traveler," the *New York Times Magazine,* March 5, 1995.

Portions of "In the Belly of the Beast" appeared in the *Tucson Weekly,* July 2, 1986.

"Jabberwocky" is adapted from an address to the American Booksellers' Convention, 1993, and several other lectures.

"The Forest in the Seeds" appeared in different form in *Natural History,* October 1993.

"Careful What You Let in the Door" is adapted from an address given as part of the San Francisco Arts & Lectures series, 1993.

"The Not-So-Deadly Sin" was published in *Waterstone's Writers Diary,* London, 1995.